Calculations for HEALTH and SOCIAL CARE

Gordon E. Gee

Formerly Head of Department of Food, Fashion and Social Services,
South Downs College of Further Education, Havant, Hants

Formerly Examiner in Book-keeping and Food Costing for the
East Midlands Educational Union

Hodder & Stoughton

A MEMBER OF THE HODDER HEADLINE GROUP

Acknowledgements

Grateful acknowledgement is due to the following persons who have provided help and technical information in the writing of this book.

Helon Bent – Planning and Development Officer, Social Services; Michael Gee – Consultant, Computergramme, Bristol; Graeme Kennedy – Senior Pharmacist, St Richard's Hospital, Chichester; Gill Kester – Lecturer, West Sussex College of Nursing and Midwifery; Ann Marchand – Hospital Sister, The Arundel and District Hospital; Janet Miller – Nurse Tutor, West Sussex College of Nursing and Midwifery; Sue Monday – Head of Community Studies Unit, Chichester College of Technology; Karen Patricia Neal – Occupational Therapy, Stoke Park Hospital, Bristol; Jenny Page – Clinical Specialist, Maternity Unit, St Richard's Hospital, Chichester; Zena Payne – Supervisor, Poppins Day Nursery, Southbourne; Lorraine Thomas – Diabetic Nurse Specialist, St Richard's Hospital, Chichester.

Cataloguing in Publication Data is available from the British Library

ISBN 0 340 60154 X

First published 1994
Impression number   10   9   8   7   6   5   4   3   2
Year   1998   1997   1996   1995   1994

Copyright © 1994 Gordon E Gee

All rights reserved. No part of this publication may be reproduced or transmitted in any form or by any means, electronic or mechanical, including photocopy, recording, or any information storage and retrieval system, without permission in writing from the publisher or under licence from the Copyright Licensing Agency Limited. Further details of such licences (for reprographic reproduction) may be obtained from the Copyright Licensing Agency Limited, of 90 Tottenham Court Road, London W1P 9HE.

Typeset by Wearset, Boldon, Tyne and Wear.
Printed in Great Britain for Hodder & Stoughton Educational, a division of Hodder Headline Plc, 338 Euston Road, London NW1 3BH by Athenaeum Press Ltd.

Preface

This book is designed for students who have chosen to enter the Health and Social Care professions. In many cases they will already be working in this vital sector, taking refresher courses or gaining further qualifications.

Cognisance of the latest GNVQ core skills has guided the author in the selection and grading of topics, together with the needs of the professions.

Health and Social Care is a wide and varied subject, but there are many areas of the work which overlap or intersect. It is therefore possible to produce a book sufficiently broad yet specific to certain requirements.

Financial considerations, more than ever before, mean that there is a need for those students seeking to gain promotion to understand basic budgeting, percentage returns, statistics, costing and the keeping of accounts.

There are one or two chapters in this book which are particularly relevant to nursing, but it is hoped that all students will find the information contained useful as exercises in arithmetical practice.

Where topics are considered worthy of further discussion, they are indicated so that tutors can explore them in depth.

The tests at the end of the book have been carefully graded to allow students to achieve a degree of success commensurate with their ability.

Contents

1	Addition	1
2	Subtraction	3
3	Multiplication	4
4	Division	5
5	Metric and Imperial measurement	8
6	Problems using the four rules	10
7	Area	12
8	Averages – mean, median and mode	15
9	Days inclusive	17
10	Twenty-four hour clock	19
11	Accounting for cash	20
12	Using fractions	24
13	Infusion rates	28
14	Ratio	30
15	Percentages – expressing one amount as a percentage of another	33
16	Percentages, fractions and ratios	35
17	Solution strength	36
18	Percentage problems	38
19	Puzzles	40
20	More about percentages – multiplying by a percentage	42
21	More percentage problems	43
22	Profit and loss	44
23	Costing a meal	46
24	Preparing a budget	49
25	Stock control	51
26	Graphs	53
27	Pie charts	57
28	Centigrade and fahrenheit	58
29	Computer spreadsheets	60
30	Mixed tests	63
	Answers	82

— 1 —
Addition

Calculating the cost of running a playgroup, nursing home, hospital, or preparing a budget for the provision of social services, depends on a capability to provide meaningful information. Accuracy in addition of numbers is therefore essential.

Add the following:

	1		2		3		4	
	425		649		85		3507
	138		288		469		934
	207		173		299		218

	5		6		7		8	
	460		4329		4023		3529
	85		787		75		6358
	1908		1376		1802		2756

Calculating duty rosters often requires working in hours and minutes.

9	hrs	mins	10	hrs	mins	11	hrs	mins	12	hrs	mins
	6	10		3	13		12	17		3	56
	2	25		5	42		0	47		2	42
	8	18		1	05		12	25		1	55

The simplest use of decimals is in money calculations, but they are also extensively used in calculating medicinal dosages and mixtures involving weights and liquids.

Remember, the first place behind the decimal place is the tenths column, the second is the hundredths, the third is the thousandths, etc. For example:

> 31.476 is 31 whole numbers, 4 tenths, 7 hundredths and 6 thousandths.

Because there are 100 pence in £1, then 1 penny is a hundredth of £1 and is written as £0.01.

13 Write 9p as a decimal of £1.
14 Write 28p as a decimal of £1.

> 1000 grammes (g) = 1 kilogramme (kg)

15 Write one gramme as one thousandth of a kilogramme.
16 Write 25 g as a decimal of 1 kg.
17 Express 856 g as a decimal of 1 kg.

1000 millilitres (ml) = 1 litre

18 Write 5 ml as a decimal of 1 litre.
19 Express 755 ml as a decimal of 1 litre.

1000 millimetres (mm) = 1 metre (m)

20 Write 38 mm as a decimal of 1 m.
21 Express 485 mm as a decimal of 1 m.

16 kg 255 g can be written as 16.255 kg
4 litres 750 ml can be written as 4.750 litres or 4.75 litres

22 Write the following in kilogrammes:
 a 12 kg 235 g
 b 7 kg 30 g
23 Write the following in litres:
 a 4 litres 50 ml
 b 6 litres 5 ml

Care must be taken when adding the following to keep the decimal points and the tenths, hundredths and thousandths in the same vertical column.

	24 £	**25** £	**26** kg	**27** kg
	171.39	3.20	3.015	0.15
	3.87	15.07	0.258	2.505
	18.29	325.76	1.35	8.044

	28 litres	**29** litres	**30** m	**31** m
	10.005	0.08	150.25	0.255
	4.226	21.455	110.384	0.765
	8.7	1.75	302.005	0.488

Sometimes it may be necessary to add horizontal numbers. Be certain to add the thousandths first, hundredths next, etc.

32 14.237 + 8.07 + 25.111
33 0.008 + 7.2 + 15.207
34 145.1 + 35.256 + 50.7 + 13
35 17.222 + 18.94 + 32.858 + 0.12

— 2 —
Subtraction

Gains in weight or height, overspend, a change in the number of patients, days between dates, giving change, comparison of performance – all require an ability to subtract.

Test yourself by working out the following:

1	936 −615	**2**	4562 −237	**3**	6534 −1829	**4**	1038 −78
5	£ 24.75 −3.49	**6**	£ 127.46 −85.78	**7**	£ 602.12 −413.04	**8**	£ 4206.54 −3766.28
9	kg 2.495 −0.627	**10**	kg 12.062 −9.163	**11**	litres 3.500 −3.425	**12**	litres 1.125 −0.25
13	m 154.255 −77.205	**14**	m 67.05 −38.138	**15**	hrs mins 2 23 −1 50	**16**	hrs mins 12 10 −10 36

17 Write the answer to question **9** in grammes.
18 Write the answer to question **12** in millilitres.
19 Write the answer to question **14** in millimetres.
20 £234.25 − £170.80
21 £107 − £16.37
22 3.225 kg − 1.775 kg
23 35 kg − 28.136 kg
24 2.35 litres − 1.755 litres
25 0.225 litres − 0.175 litres
26 156.258 m − 86.009 m
27 25.444 m − 24.6 m
28 Find the difference between £123.85 and £76.86.
29 What must be added to 1.255 litres to make 2 litres?
30 By how much is 17.236 metres shorter than 26.175 metres?
31 By how much is 5.45 kg heavier than 5.28 kg?
32 How much change from a £10 note should be given for purchases worth £3.33?

— 3 —
Multiplication

To be successful in this section, a sound knowledge of multiplication tables is required. How good are you?

1	4×6	**2**	5×7	**3**	3×8	**4**	6×6
5	2×12	**6**	9×4	**7**	4×11	**8**	7×3
9	6×7	**10**	8×9	**11**	12×7	**12**	7×7
13	11×11	**14**	9×12	**15**	3×9	**16**	9×6
17	12×12	**18**	9×7	**19**	5×12	**20**	11×12

Now use your knowledge of tables to work out the following:

21	39×7	**22**	135×8	**23**	306×9
24	142×35	**25**	376×48	**26**	705×96

When multiplying decimals, first work out the sum ignoring the decimal point, for example:

$$\begin{array}{r} 7.53 \times \\ 6 \\ \hline 4518 \end{array}$$

Next count the number of figures behind the decimal point in the question and place the decimal point so that there are the same number of figures behind the point in the answer:

$$\begin{array}{r} 7.53 \times \\ 6 \\ \hline 45.18 \end{array}$$

Here is another example:

$$\begin{array}{r} 3.27 \times \\ 0.4 \\ \hline 1.308 \end{array}$$

N.B. There are a total of three figures behind the decimal point in the question (2, 7, 4).

Here is an unusual example:

$$\begin{array}{r} 0.26 \times \\ 0.12 \\ \hline 0.0312 \end{array}$$

Notice how the 0 is inserted in the tenths column to give a total of four places behind the decimal point.

When using a calculator the decimal point is entered straightaway, not when the multiplication is completed.

27	14.6 × 5	**28**	2.73 × 6	**29**	36.7 × 0.4
30	£7.46 × 7	**31**	£12.56 × 12	**32**	£0.25 × 0.3
33	3.125 kg × 8	**34**	2.076 kg × 1.2	**35**	10.224 litres × 0.5
36	8.05 m × 0.04	**37**	16.25 m × 3.6	**38**	0.325 litres × 1.1
39	£4.33 × 123	**40**	2.336 m × 225	**41**	8.4 litres × 30

Quick methods

45.76 × 10 = 457.6
3.069 × 100 = 306.9
6.94 × 1000 = 6940

The decimal point is moved one place to the right when multiplying by 10, two places when multiplying by 100, three places when multiplying by 1000, and so on.

£34.45 × 10 = £345.50
2.345 kg × 100 = 234.5 kg
7.678 litres × 1000 = 7678 litres
7.5 m × 100 = 750 m
£5.24 × 20 = £104.80 (×10, then ×2)

Use the quick method to work out the following:

42	£12.34 × 10	**43**	2.355 kg × 100	**44**	18.555 litres × 200
45	1.275 m × 1000	**46**	7.5 kg × 1000	**47**	8.4 litres × 20
48	0.4 m × 300	**49**	£1.06 × 10 000	**50**	1.234 litres × 10

— 4 —

Division

Sharing, averaging, costing and many other computations depend on a competence in division. Again, this requires a knowledge of tables.

1	18 ÷ 3	**2**	45 ÷ 5	**3**	88 ÷ 8	**4**	120 ÷ 12
5	81 ÷ 9	**6**	56 ÷ 8	**7**	15 ÷ 5	**8**	72 ÷ 6
9	24 ÷ 8	**10**	14 ÷ 2	**11**	96 ÷ 12	**12**	48 ÷ 6
13	42 ÷ 7	**14**	30 ÷ 5	**15**	63 ÷ 9	**16**	132 ÷ 12

17 32 ÷ 4 **18** 60 ÷ 12 **19** 121 ÷ 11 **20** 64 ÷ 8

Using the tables:

21 102 ÷ 6 **22** 968 ÷ 8 **23** 4020 ÷ 12
24 £5915 ÷ 7 **25** 2412 kg ÷ 9 **26** 2512 m ÷ 4
27 1125 ÷ 25 **28** 5412 ÷ 41 **29** 54 384 ÷ 132

Do not forget to include the 0 in the answer to the following type of example. It does have a value.

30 1525 ÷ 5 **31** 7605 ÷ 15 **32** £24 603 ÷ 3

In decimal division, the following rules apply:

a Before working the sum out, first place the decimal point in the answer space directly above the decimal point in the question, for example:

$$\begin{array}{r} . \\ 6\overline{)45.72} \end{array}$$

Then divide:

$$\begin{array}{r} 7.62 \\ 6\overline{)45.72} \end{array}$$

33 98.4 ÷ 4 **34** 29.75 ÷ 7 **35** 110.7 ÷ 9

b Do not divide by a decimal unless a calculator is used, for example:

$$1.2\overline{)5.424}$$

Multiply both the divisor (1.2) and the dividend (5.424) by 10, and the sum can be set out as follows:

$$12\overline{)54.24}$$

The sum can now be worked out:

$$\begin{array}{r} 4.52 \\ 12\overline{)54.24} \end{array}$$

Here is another example:

$$0.25\overline{)6.375}$$

The setout and answer is:

$$\begin{array}{r} 25.5 \\ 25\overline{)637.5} \end{array}$$ (multiply divisor and dividend by 100)

N.B. The divisor and the dividend must always be multiplied by the same amount.

36 $2.76 \div 1.2$ **37** $2.52 \div 0.4$ **38** $7.775 \div 0.25$

c By adding 0 to any remainder and dividing again, a more accurate answer can be obtained, for example:

$29.7 \div 5$

The setout and answer is:

$$\begin{array}{r} 5.94 \\ 5\overline{)29.70} \end{array}$$

Depending on the accuracy required, a 0 can be added to every remainder, for example:

$$\begin{array}{r} 0.9825 \\ 4\overline{)3.9300} \end{array} \qquad \begin{array}{r} 3.1375 \\ 8\overline{)25.1000} \end{array}$$

39 $18.3 \div 5$ **40** $45.9 \div 6$ **41** $18.25 \div 4$

Sometimes an answer is required to be worked out to a given number of places after the decimal point, for example:

$92.1 \div 7$

Give the answer to 2 decimal places.

Answer:

$$\begin{array}{r} 13.15 \\ 7\overline{)92.10} \end{array}$$

Ignore any remainder after 2 decimal places.

42 $25.1 \div 8$ (answer to 2 decimal places)
43 $16.23 \div 7$ (answer to 3 decimal places)
44 $123.4 \div 6$ (answer to 2 decimal places

Approximating an answer by *correcting* to a given number of decimal places is a useful function.

Example

Express 13.357 corrected to 2 decimal places.

Method

Look at the figure in the third decimal place. If it is 5 or of a greater value then add 1 to the figure in the second decimal place. If the value is less than 5 then leave the figure in the second place unchanged.

13.357 corrected to 2 decimal places is 13.36

N.B. Check the figure in the second place if your are correcting to

1 decimal place, check the third place if you are correcting to 2 decimal places, and so on.

Here are more examples:

>130.234 corrected to 2 decimal places is 130.23
>7.0185 corrected to 3 decimal places is 7.019
>£25.236 corrected to 2 decimal places is £25.24
>£38.517 to the nearest penny is £38.52

45 Express 2.238 corrected to 2 decimal places.
46 Express 145.4442 corrected to 3 decimal places.
47 Express 17.05 corrected to 1 decimal place.
48 Express £203.897 to the nearest penny (corrected to 2 decimal places).

Quick methods

>$504.3 \div 10 = 50.43$
>$1425.6 \div 100 = 14.256$
>$18.45 \div 1000 = 0.01845$

The decimal point is moved one place to the left when dividing by 10, two places when dividing by 100, three places when dividing by 1000, and so on.

49 $308.2 \div 10$ **50** $1025.3 \div 1000$ **51** $£42.00 \div 100$
52 $565.5 \text{ kg} \div 10$ **53** $8.255 \text{ m} \div 100$ **54** $14 \text{ litres} \div 10$

— 5 —
Metric and Imperial measurement

Although the government signalled an intention in 1971 to convert from an Imperial to a Metric system of measurement, the plan hasn't been fully carried out.

Imperial measurement is still used in some areas of commerce and is commonly used by most elderly people. To tell grandparents that their newborn grandchild weighs 3.4 kilogrammes is usually meaningless to them, but they readily understand and relate to $7\frac{1}{2}$ pounds. Similarly, to many older persons a height of 6 feet is more easily visualised than 1.83 metres. Despite the introduction of Metric measurement, we still order a pint of beer, a pound of tomatoes, and sometimes a square yard of carpet.

As the Care industry is primarily concerned with people, students should be aware of the abilities and limitations of their charges. Remember, differing abilities are not lesser abilities.

Commonly used measurements

Weight

Metric	Imperial
1000 grammes (g) = 1 kilogramme (kg)	16 ounces (oz) = 1 pound (lb)
1000 kg = 1 tonne	14 lbs = 1 stone
	112 lbs = 1 hundredweight (cwt)
	20 cwts = 1 ton

1 kg is approximately 2.2 lbs

Further Metric weight tables can be found on page 10.

Length

1000 millimetres (mm) = 1 metre (m)   12 inches (in) = 1 foot (ft)
1000 m = 1 kilometre (km)   3 ft = 1 yard
    1760 yards = 1 mile

1 m is approximately 39.4 inches
1 km is approximately 0.62 miles
1 mile is approximately 1.61 km

Capacity

1000 millilitres (ml) = 1 litre   8 pints = 1 gallon

1 litre is approximately 1.8 pints

Example

Express 20 kg as lbs.

Method

$$1 \text{ kg} = 2.2 \text{ lbs}$$
$$\text{Therefore } 20 \text{ kg} = 20 \times 2.2 \text{ lbs}$$
$$= 44 \text{ lbs (approx.)}$$

Example

Express 10 pints as litres.

Method

$$1 \text{ litre} = 1.8 \text{ pints}$$
$$\text{Therefore } 10 \text{ pints} = 10 \div 1.8 \text{ litres}$$
$$= 5.555 \text{ litres (approx.)}$$

1 Express 30 kg as lbs.
2 Express 25 pints as litres.
3 Express 50 oz as lbs and oz.

4 Express 100 cwt as tons.
5 Express 100 miles as km.
6 Express 110 lbs as kg.

Further Metric tables which are useful for students of nursing:

1000 microgrammes = 1 milligramme (mg)
1000 milligrammes = 1 gramme (g)

7 Express 3500 microgrammes as: **a** milligrammes, **b** grammes.
8 Express 46 milligrammes as: **a** grammes, **b** microgrammes.
9 Express 1 gramme as: **a** milligrammes, **b** microgrammes.
10 Express 1 kilogramme as milligrammes.
11 How many tablets, each of 25 milligrammes, are necessary to give a dose of 0.075 grammes?
12 A tablet contains 100 microgrammes. How many tablets will provide 0.25 milligrammes?
13 A tablet contains 250 milligrammes of soluble aspirin. What dosage would provide 125 milligrammes?

— 6 —
Problems using the four rules

1 The Tiny Tots Day Nursery charge £6.36 per session, £1.85 for a cooked lunch and £1.38 for a tea. Calculate the total cost to Mrs Wilson in a four week month if each week her son attends the nursery for two mornings and three afternoon sessions, and has two lunches and one tea.
2 After taking the measurement of Mr Reed's leg muscle, a nurse found there had been a reduction from 0.39 m to 0.378 m. Express the difference in millimetres.
3 In using her own car for casework, a midwife is allowed to claim £0.46 per mile. At the start of one day the mileometer in her car registered 025957, but by the end of her shift the mileometer showed 026011. Calculate how much the midwife should claim for the day.
4 A child is prescribed a drug at the rate of 35 mg for each kg of

weight (per dose). Calculate how many mg are required for each dose if the weight of the child is 18 kg.

5 A newborn baby weighed 3.525 kg. The baby's grandmother wanted to know the weight expressed in lbs.
   a Give the weight as requested by the grandmother.
   b Is the weight approximately: (i) $7\frac{1}{2}$ lbs, (ii) $7\frac{3}{4}$ lbs, (iii) 8 lbs?

6 A hospital chef was allowed to spend £1.78 per patient for the evening meal. If in preparing the meal for 84 patients he spent £162.25, calculate the difference in total between the amount allowed and the actual cost, and say whether it was an overspend or underspend.

7 How many 40 mg tablets should be given for a dose of 0.1 grammes?

8 A concentrated stock solution has to be diluted with distilled water to bring it to the correct strength. If 5 litres of solution are produced by using 425 ml of stock:
   a how many ml of distilled water must be added?
   b Express this answer in litres.

9 A Health Authority post for a Nursing Assistant Grade A was advertised at £7000 per annum and involved a 37.5-hour week. John considered applying for this position, but he was already employed for 37 hours per week in a private nursing home which paid £4.10 per hour. Assuming a 52-week year for the purposes of this question:
   a which was the better paid job annually?
   b which job paid the most per hour and by how much?

10 Sarah, a dental receptionist, worked the following hours in one week:

   Monday   8.30 a.m. – 12.30 p.m.   2.00 p.m. – 5.00 p.m.
   Tuesday   8.30 a.m. – 1.00 p.m.   2.00 p.m. – 5.15 p.m.
   Wednesday 8.00 a.m. – 12.30 p.m.   1.30 p.m. – 5.15 p.m.
   Thursday   8.00 a.m. – 12.30 p.m.   2.00 p.m. – 5.00 p.m.
   Friday   8.00 a.m. – 12.30 p.m.   2.00 p.m. – 5.00 p.m.

   If Sarah was paid £4.60 per hour for a 37-hour week plus time and a half for any overtime worked, calculate the wages she received for the week.

11 The catering section of the Brierly Engineering Company had spare kitchen capacity and agreed to provide 105 cooked meals twice a week for the Meals on Wheels service. The total charge for this provision was £157.50 per week. What was the cost per meal?

12 A lady presented a chemist with a doctor's prescription for three

separate items each costing £4.75. Find the change she received if she tendered a £20 note.

13 A diet for newly diagnosed diabetic children recommended the following daily carbohydrate content: Ten times 10 g portions for the first year of life plus 10 g extra for each additional year of life until the age of twelve. Calculate the daily carbohydrate intake laid down for a nine year old.

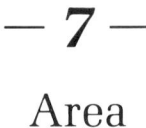

Area

Day nurseries have to obey regulations regarding area of floor space, and most business premises are taxed according to floor area. There is a need, therefore, to be able to calculate an area accurately.

Example

Calculate the area of the floor space shown below.

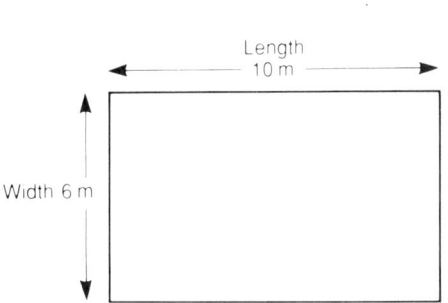

Method

The area is found by using the formula:

Area = Length × Width (Breadth)
Area = 10 × 6 = 60 square metres

1 Find the area of a room 12 m long and 9 m wide.

2 How many square feet are contained in an area measuring 20 ft by 15 ft?

3 Allowing for a cupboard of length 3 m and width 1.5 m, calculate the area of the room for carpeting purposes if the overall measurements are length 14 m, width 12 m.

4 Local regulations stipulated the following minimum areas per child for day nurseries:

up to 1 year old – 40 square feet
between 1 and 2 years old – 30 square feet
between 2 and 3 years old – 20 square feet

The plan (continuous line) shown below represents the two rooms hoping to be used by the Littletown Day Nursery.

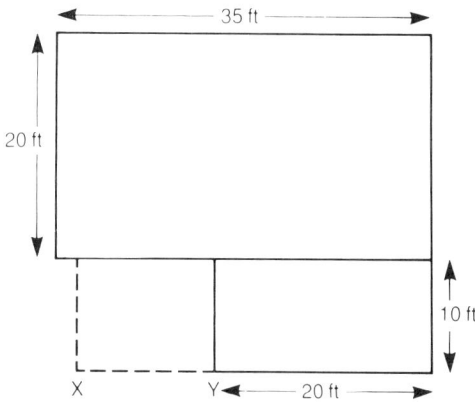

It was proposed to limit the intake of children to:

up to 1 year old – 6 children
between 1 and 2 years old – 18 children
between 2 and 3 years old – 10 children

a Would it be allowable to use the rooms as shown? (Ignore the effect of toilet requirements etc.)
b If it was possible to expand along the line towards X, what is the minimum length of X–Y to allow for the proposed intake?

5 Calculate the cost of covering the floor of the room shown overleaf with carpet 4 m wide, if the carpet was priced at £13.65 per square metre and could only be purchased by the metre (i.e. parts of a metre could not be bought).

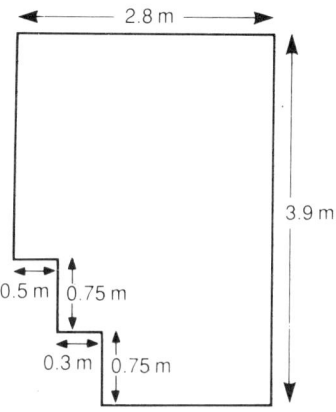

A common error is over the use of the term 'square'. Study the diagram below.

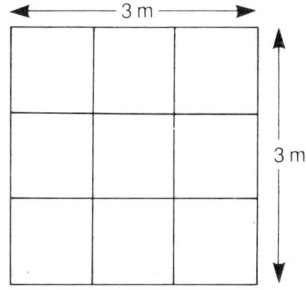

An area 3 m by 3 m has an area of 9 square metres, but it can also be referred to as a '3 metre square'.

6 How many square metres are contained in a 2 metre square?

7 A room is 5 metres square. Calculate the area of the room in square metres.

8 Find the cost of carpeting a 4 yard square cubicle if the cost of the carpet was £12.50 per square yard.

9 What is the difference in square feet between areas 10 square feet and 10 feet square?

— 8 —
Averages – mean, median and mode

Finding the mean *average* patient length of stay in hospital may be useful from a costing point of view, but it may not give a *typical* length of stay or even an *actual* length of stay, for example:

In a ward during one week, nine patients stayed for the following number of days:

    5, 7, 30, 7, 4, 3, 7, 4, 5

The **mean** average stay is found by adding the total number of days and dividing by the number of patients:

    72 (days) ÷ 9 (patients) = 8 days

However, no patient actually stayed for eight days and the average is distorted due to the abnormally high stay of 30 days by one patient.

In this case the **median** average would be more useful for showing the typical stay. This is found by arranging the days in order of value and then finding the middle value:

    3, 4, 4, 5, $\underline{5}$, 7, 7, 7, 30

The median is 5 and the distribution will be spread equally above and below that figure.

If there are an even number of patients staying, e.g. for 2, 4, 6 and 7 days, the median is the mean average value of the middle two figures:

    (4 + 6) ÷ 2 = 5 days

Another useful value is the **mode**, which is found by examining the distribution of the days to find which number occurs more frequently.

By studying our example you will find the frequencies as follows:

    3 days   once
    4 days   twice
    5 days   twice
    7 days   three times
   30 days   once

The most frequent occurrence is 7 days and that is said to be the mode.

It is important when using these terms to understand their

significance. In the example given, the arithmetically exact average was 8 days, a sensible informative average was 5 days and the most common length of stay was 7 days. Their use depends on which information is required and for what purpose.

1. In the previous example involving the nine patients:
   a calculate the total cost using the arithmetical mean figure of 8 days, if the cost per patient per day was £150.
   b On the evidence presented, what is the most common length of stay in the ward?

2. In a study of ten men, each 1.775 m in height, the following weights were recorded:

   70 kg, 71.2 kg, 65.5 kg, 75.25 kg, 70.5 kg, 73 kg, 75.25 kg, 77 kg, 67.6 kg, 75.25 kg

   a Calculate the average (arithmetical mean) weight of the men.
   b Find the modal weight.

3. During a year, the average number of rooms occupied each night at the Grange Private Hospital was 14. If the charge per room per night was £200, calculate the total room receipts for the year.

4. In planning an appointments book for a doctor's surgery, past records of 250 patients showed the lengths of consultations to have been as follows:

No. of patients	Length of consultation
16	8 minutes
46	7 minutes
80	5 minutes
108	10 minutes

   a Would the best method of arranging future consultations be based on finding the mean, median or mode of the above details?
   b What is the average time per consultation? (Answer to the nearest minute.)

5. A dentist's receptionist was asked to prepare a survey of children's teeth by selecting ten patients based on certain criteria, to find the most common number of fillings per child. Give the figure the receptionist should submit, based on the information below.

Patient	No. of fillings
A	3
B	4
C	3
D	2
E	4

F	4
G	1
H	9
I	6
J	4

6 An analysis of the times taken in minutes by an ambulance crew to reach reported incidents is reproduced below.

5, 6, 7, 10, 6, 5, 10, 10, 12, 6, 40, 7, 10, 12, 10, 6, 10, 6, 7, 10.

For the purposes of this question, assume the distance travelled on each occasion was similar.
    a Calculate the arithmetical mean response time.
    b Is the answer to **a** a fair method of producing a statistic?
    c What other method would be appropriate in this case?

— 9 —

Days inclusive

When finding the length of stay in hospital, the period of treatment, staff leave, absence due to sickness, deterioration of stock, etc., it is essential that the precise number of days can be calculated.

If a patient was admitted to hospital Monday morning, 12 January and was discharged Monday afternoon, 19 January, then depending on the method of accounting for bed occupation, it could be said that the patient stayed for seven nights or eight days inclusive. Certainly it could be shown that eight lunches were consumed – if meals were not restricted medically.

Example

A member of staff is absent from work from 13 February until 17 February inclusive (including both dates). How many days absence should be recorded?

Method

        17 − 13 = 4 (subtraction of dates)
        4 + 1 = 5 (inclusive of 17 February)

Proof

Days in February involved – 13, 14, 15, 16, 17 = 5 days

Rule

When calculating the number of days inclusive, find the difference between the dates and add one.

1. Find the number of days inclusive between 2 April and 27 April.

2. A group of students on a Community Care course needed insurance for work placement. How many days cover should be arranged if they commenced work on 4 May and finished on 17 May in the same year.

When calculating days involving two or more months, the additional day is only added to the first month.

Example

Find the number of days inclusive between 26 September and 7 November in the same year.

Method

    Days in September =   5 (30 – 26 + 1)
    Days in October   = 31
    Days in November   = 7
    Days in total   = 43

3. A playgroup helper phoned on 26 January to say she had 'flu' and wouldn't be available. She returned to work on 2 February. For how many working days was the helper absent if two of the days she was ill involved a weekend when the playgroup did not function?

4. The Toddlers Day Nursery was closed on 25 August when new safety regulations forced the building of an extension. The builder estimated that the work would be completed on 10 October and the unit could be opened again on the next day. Calculate the number of days inclusive that the nursery would be closed (include weekends in your answer).

5. For how many days was a course of treatment which commenced on 28 January and finished on 15 February of the same year?

6. On 26 May, Michael was prescribed a course of 28 tablets to be taken at the rate of two per day, starting on the day of prescription. On what date would Michael take the last tablet?

— 10 —
Twenty-four hour clock

In most cases meetings and appointments are arranged according to the twelve hour clock, but occasionally the twenty-four hour system is used to minimize mistakes in fixing a time, especially when logging a call-out for the emergency services.

Here are a few examples showing the same time for the twelve hour and the twenty-four hour systems:

12 Hour	24 Hour
7 a.m.	07.00
10.30 a.m.	10.30
3 p.m.	15.00
11.15 p.m.	23.15

The twenty-four hour system measures the time from midnight and uses two figures for hours and two figures for minutes. Sometimes the word 'hours' is written after the figures, e.g. 13.30 hours.

Express the following times using the twenty-four hour system:

1. 3 a.m.
2. 1 p.m.
3. 9.45 a.m.
4. 6.15 p.m.
5. 11.59 p.m.
6. 12.05 a.m.
7. midday
8. 9.24 p.m.

Express the following times using the twelve hour system:

9. 11.00
10. 14.00
11. 21.35
12. 00.20
13. 05.25
14. 12.10
15. 22.00
16. 01.00

17. A call-out was logged in at 02.48 and a vehicle arrived at the incident at 03.07. Find the total response time.

18. Find the length of time taken for a journey which commenced at 11.53 and ended at 14.09.

19. An appointment was made for a lady to see a doctor at 3.20 p.m., but due to an emergency the doctor's appointments were running three-quarters of an hour late. Give the time the lady should see the doctor.

20. If a nurse's shift started at 7.30 a.m. and finished at 3.15 p.m., calculate the length of her working day.

21. During a night, a patient was visited at 22.40 and again at 00.55. Find the interval between visits.

22 A witness claimed that he called an emergency service at 'ten to nine' in the evening, but his call was logged at 21.03. Find the difference in minutes between the conflicting times.

− 11 −
Accounting for cash

The Health and Social Care industry is regularly under scrutiny wherever money is concerned. Apart from the multi-million pounds of tax payers' money involved in the running of such an immense organisation, employees are often required to handle and account for the relatively small yet vital amounts of cash belonging to patients in their care.

In this chapter, a simple method of accounting for income and expenditure will be shown. It is suitable for a small business such as a playgroup, or for keeping a record of a patient's spending money. It is recommended that accounts of all transactions made on behalf of patients are kept up to date, both for the benefit of the patient and the protection of the carer. Invoices and receipts should be retained for future reference and for reclaiming Value Added Tax where allowable.

Set out below is a simple cash book showing money received on the left hand side and money paid out on the right hand side.

Income				**Expenditure**		
Date			£	*Date*		£
1 Jan	Balance b/d		100.00	7 Jan	Window cleaner	20.00
12	Cash received		800.00	10	Chair	35.00
				12	Food and drink	76.00
				14	Petrol	22.50

N.B. The balance of £100.00 was brought down from 31 December.

Now to balance the account on 14 January:

Income			Expenditure		
Date		£	*Date*		£
1 Jan	Balance b/d	100.00	7 Jan	Window cleaner	20.00
12	Cash received	800.00	10	Chair	35.00
			12	Food and drink	76.00
			14	Petrol	22.50
			14	Balance c/d	746.50
		900.00			900.00
15 Jan	Balance b/d	746.50			

Method

    a Add the largest value side (in this case the left – £900).

    b Total the expenditure – £153.50.

    c Subtract the £153.50 from £900.00 to find the balancing amount – £746.50, and enter that figure under the expenditure.

    d Enter the left and right side totals – £900.00, and draw lines as indicated.

N.B. Both sides must agree in total.

Notes

In the above example, the balance to be carried down (c/d) has been brought down (b/d) to start the next accounting period on 15 January.

The length of the accounting period can vary according to requirements.

For security, do not leave a space between the balance c/d and the last item of expenditure, i.e. £22.50. This prevents insertions of further expenditure after the account is balanced.

If the account is overspent, then the balance c/d will appear on the left and the new balance b/d will be entered on the right.

1 In the above example:
   a by how much did the income exceed the expenditure over the two weeks?
   b how much money was left in the account to commence the new accounting period on 15 January?

2 Use the method above to keep a personal account for a special needs patient, given the following information:

    1 February – Balance brought down from previous week £10.00
    1 February – Received cash £15.00

1 February – Purchased chocolate £1.75
3 February – Travel cost £2.00
3 February – Entrance fee to skating £3.50
4 February – Cost of pub lunch £4.25
5 February – Balance account
8 February – Bring balance down to start new period

3. The cash book of the Knee High Playschool showed a balance brought down of £154.78 to start the week on Monday 14 March. Complete the account for the week from the following data:

14 March – Purchased cupboard £36.00
15 March – Purchased toys £16.50
15 March – Purchased drinks £6.90
17 March – Received gift of £50.00
17 March – Paid electricity £91.75
18 March – Balance account

Petty cash

Sometimes in a large organisation a person is required as part of his/her job to keep an account of small amounts of expenditure. Details of this expenditure are usually recorded in an *imprest petty cash book*. The setout is similar to the cash book.

At the beginning of an accounting period, an agreed sum of money called 'the float' is given to the responsible person who then keeps an account of any money spent.

At the end of the period, any expenditure is then claimed back to make up the float.

The amount of float is determined by past records of expenditure.

Here is an example of a petty cash book (account):

Income			**Expenditure**		
Date		£	*Date*		£
1 May	Cash	20.00	2 May	Stamps	7.50
			4	Travel	2.25
			4	Pencils	1.14
			5	Tape	6.00

N.B. the float is £20.00.

The left hand side shows income received. The right hand side shows all expenditure.

The book is to be balanced on 6 May:

Income			**Expenditure**		
Date		£	*Date*		£
1 May	Cash	20.00	2 May	Stamps	7.50
			4	Travel	2.25
			4	Pencils	1.14
			5	Tape	6.00
			6	Balance c/d	3.11
		20.00			20.00
8 May	Balance b/d	3.11			
8	Cash	16.89			

Method

    a Balance the account as a cash book.
    b Having brought down the balance of £3.11, then claim the expenditure of £16.89 to make up the float to £20.00.
    c Verify expenditure with receipts.

4 The float of the petty cash book maintained by the Young Set Day Nursery was set at £25.00. Complete the account from the following details:

    8 January – Received the full float
    9 January – Purchased Sellotape £0.75
    9 January – Purchased diary £2.50
    10 January – Purchased stamps £2.80
    11 January – Purchased first aid kit £6.70
    11 January – Taxi £8.00
    12 January – Purchased staples £1.25
    13 January – Balance account and carry down
    15 January – Received cash to make up float

5 The petty cash book of the Baybourne Surgery had a balance brought down on 17 May of £8.15. The float was maintained at £30.00. Enter the following transactions for the week commencing 17 May:

    17 May – Draw cash to make up float
    17 May – Purchased appointments book £8.25
    18 May – Purchased clock battery £2.35
    18 May – Purchased ribbons for printer £8.05
    20 May – Purchased magazine £1.70
    21 May – Purchased stationery £6.88
    21 May – Balance account

— 12 —
Using fractions

In the following chapters think of a fraction as a division sum, e.g. $\frac{1}{2}$ is $1 \div 2$, $\frac{3}{4}$ is $3 \div 4$.

This will help to minimize difficulty in setting out sharing and percentage calculations, and in expressing vulgar fractions as decimal fractions.

Cancelling

Cancelling makes the working out of fractions easier by simplifying the numbers used. For example, two-quarters are usually referred to as one-half.

$$\frac{2}{4} = \frac{1}{2}$$

The value remains the same because both the top and the bottom lines are divided by two.

The secret is to find a number that will divide exactly into both lines.

$\frac{35}{40} = \frac{7}{8}$ (by dividing both lines by 5)

Cancel the following. (Always examine your answer to see if it will cancel again.)

1. $\frac{12}{15}$
2. $\frac{9}{12}$
3. $\frac{20}{35}$
4. $\frac{36}{42}$
5. $\frac{16}{64}$
6. $\frac{100}{1000}$
7. $\frac{17}{51}$
8. $\frac{14}{42}$
9. Express £1 as a fraction of £3.
10. Express £26 as a fraction of £39 (cancel if possible).
11. A survey showed that 17 out of 68 students questioned regularly engaged in physical exercise. Express this finding as a fraction cancelled down to give a more easily understood statistic.
12. A lotion mixed as 1 part stock and 60 parts water is to be prepared. Express the stock as a fraction of: **a** the water, **b** the mixture.
13. In producing a lotion, 15 parts stock were used with 80 parts distilled water.
    a Express the stock as a fraction of the water.
    b Cancel the answer if possible.
    c Express, in the lowest terms, the stock as a fraction of the lotion.

Mixed numbers and improper fractions

An amount containing a whole number and a fraction is known as a mixed number, e.g. $4\frac{3}{4}$.

A fraction with the value of the top line (numerator) greater than the bottom line (denominator) is known as an improper fraction, e.g. $\frac{14}{5}$.

It is important to be able to deal with these fractions with confidence.

Example

Change $5\frac{2}{3}$ to thirds.

Method

In 5 whole numbers there are 15 thirds (5×3)
Therefore in $5\frac{2}{3}$ there are in total 17 thirds $= \frac{17}{3}$

The mixed number $5\frac{2}{3}$ has been expressed as an improper fraction, $\frac{17}{3}$.

14 Change $2\frac{1}{2}$ to halves
15 Change $3\frac{4}{5}$ to fifths
16 Change $3\frac{3}{4}$ to quarters
17 Change $7\frac{4}{9}$ to ninths
18 Change $10\frac{7}{8}$ to eighths
19 Change $6\frac{11}{12}$ to twelfths
20 Change $1\frac{19}{25}$ to twenty-fifths
21 Change $40\frac{1}{3}$ to thirds

Example

Change $\frac{29}{6}$ to a mixed number.

Method

The fraction $\frac{29}{6}$ means $29 \div 6$
$29 \div 6 = 4$ whole numbers and a remainder of $\frac{5}{6}$
Therefore $\frac{29}{6} = 4\frac{5}{6}$

Change the following to mixed numbers:

22 $\frac{22}{7}$
23 $\frac{16}{5}$
24 $\frac{41}{3}$
25 $\frac{109}{12}$
26 $\frac{35}{8}$
27 $\frac{17}{2}$
28 $\frac{44}{8}$ (cancel)
29 $\frac{51}{6}$
30 $\frac{75}{20}$

Multiplication of fractions

Unless you are using a calculator, it is impossible to work out problems involving percentages without a knowledge of fraction multiplication.

$\frac{2}{3}$ of $\frac{3}{4}$ can be expressed as $\frac{2}{3} \times \frac{3}{4}$

Set out the following as fraction multiplications, but do not work the sums out:

31 $\frac{3}{4}$ of $\frac{4}{5}$ 32 $\frac{2}{7}$ of $\frac{5}{11}$ 33 $\frac{15}{17}$ of $\frac{21}{50}$

Example a

$\frac{2}{3} \times \frac{4}{7}$

Method

Multiply the top and bottom lines:

$\frac{2}{3} \times \frac{4}{7} = \frac{8}{21}$

Example b

$\frac{3}{4} \times \frac{5}{6}$

Method

If possible, cancel any figure on the top line with any figure on the bottom line to make the sum easier:

$\frac{\cancel{3}^1}{4} \times \frac{5}{\cancel{6}_2} = \frac{5}{8}$

Example c

$2\frac{1}{2} \times \frac{3}{4}$

Method

All mixed numbers must be changed to improper fractions before multiplying:

$\frac{5}{2} \times \frac{3}{4} = \frac{15}{8} = 1\frac{7}{8}$

Multiply the following:

34 $\frac{1}{2} \times \frac{3}{5}$ 35 $\frac{2}{7} \times \frac{3}{5}$
36 $\frac{8}{9} \times \frac{2}{5}$ 37 $\frac{9}{11} \times \frac{2}{3}$
38 $\frac{4}{7} \times \frac{1}{2}$ 39 $\frac{5}{12} \times \frac{7}{20}$
40 $\frac{4}{9} \times \frac{15}{16}$ 41 $\frac{3}{5} \times \frac{25}{36}$
42 $\frac{2}{3}$ of $\frac{4}{7}$ 43 $\frac{4}{5}$ of $\frac{5}{8}$
44 $1\frac{1}{2} \times \frac{4}{5}$ 45 $3\frac{3}{4} \times \frac{3}{5}$
46 $5\frac{1}{3} \times 1\frac{1}{4}$ 47 $5\frac{1}{4} \times 3\frac{1}{7}$
48 $3\frac{1}{2} \times 3$ (write as $\frac{3}{1}$)   49 $2\frac{2}{5} \times 5$
50 Find three-quarters of two-thirds.
51 Find two and a half times one-half.
52 Find one-hundredth of $\frac{4}{5}$.
53 Find $\frac{3}{100}$ of £1.00 (change to pence).

Division of fractions

Although not so commonly used as multiplication, certain calculations, such as finding the correct drug dosage, require a knowledge of fraction division.

Example

$\frac{3}{4} \div \frac{2}{3}$

Method

The sum is set out as a multiplication but the divisor ($\frac{2}{3}$) is inverted:

$\frac{3}{4} \times \frac{3}{2} = \frac{9}{8} = 1\frac{1}{8}$

54   $\frac{2}{3} \div \frac{5}{7}$ 55 $\frac{3}{4} \div \frac{5}{8}$

56   $2\frac{1}{2} \div \frac{5}{6}$ 57 $\frac{1}{2} \div \frac{1}{4}$

58   $3\frac{2}{3} \div 1\frac{1}{2}$ 59 $4\frac{2}{3} \div 1\frac{3}{4}$

60   one-half ÷ one-half   61   two-thirds ÷ one and a half

Expressing a vulgar fraction as a decimal fraction

Remember, a fraction is a convenient way of setting out a division sum, e.g. $\frac{1}{4} = 1 \div 4$.

$$\frac{1}{4} = \frac{0.25}{4\overline{)1.00}} \text{ therefore } \frac{1}{4} = 0.25$$

Simply divide the top line (numerator) by the bottom line (denominator).

62   Express $\frac{1}{2}$ as a decimal.   63   Express $\frac{2}{5}$ as a decimal.

64   Express $\frac{3}{8}$ as a decimal.   65   Express $\frac{7}{20}$ as a decimal.

66   If three out of every five students were female, express this statistic as a decimal of the total student population.

67   If five out of every 30 children examined needed to wear reading spectacles, express this fact as a decimal of the group examined.

— 13 —
Infusion rates

Nurses are regularly required to set and check the rate of drip of an Intravenous Set. To carry out this extremely important function, a knowledge of metric weights and capacity tables is needed, together with confidence in multiplication and division of decimal fractions.

The rate of drip to be set depends on the amount of solution required, the number of drops to provide that solution, and the length of time specified.

Provided the number of drops required to produce 1 ml of solution is known, then a simple multiplication will give the number of drops for any amount of solution.

IV sets in general use give 15 drops per ml, 20 drops per ml, or 60 drops per ml (used in Paediatrics).

Example

An IV set delivers 20 drops per ml. How many drops will give 0.5 ml of solution?

Method

Drops required = $0.5 \times 20 = 10$
or simply $\frac{1}{2}$ of $20 = 10$

Remember, 'of' means multiply.

1. If 20 drops deliver 1 ml of solution, how many drops will deliver: **a** 0.1 ml, **b** 0.4 ml, **c** 0.05 ml, **d** 5 ml, **e** 20 ml, **f** 100 ml, **g** 500 ml.

Example

An IV set delivers 20 drops per ml. If 0.5 ml of solution must be given every minute (0.5 ml/min), calculate the drip rate.

Method

Solution required per minute = 0.5 ml
But 20 drops deliver 1 ml
Therefore drops required per minute = $0.5 \times 20 = 10$
Drip rate = 10 drops per minute

2. An IV set delivers 20 drops per ml. Find the drip rate required to produce: **a** 0.7 ml per minute, **b** 0.8 ml/min, **c** 0.34 ml/min. (Give your answer to the next highest whole number.)

Example

100 ml of fluid is required over one hour. Find the drip rate per minute required using an IV set delivering 20 drops per ml.

Method

    Fluid required per hour = 100 ml
    Drops required per hour = 100 × 20 = 2000
    Drops required per minute = 2000 ÷ 60 = 33.33
    Drip rate = 34 (to the next highest whole number)

3. Using an IV set delivering 20 drops per ml, find the drip rate required to produce: **a** 80 ml over one hour, **b** 100 ml over two hours, **c** 500 ml over four hours.

4. A transfusion of 360 ml of blood is to be given over three hours. Assuming an IV set rate of 15 drops per ml, calculate the drip rate required in drops per minute.

5. A child is prescribed 200 ml of solution over four hours. The IV set delivers 60 drops per ml. Find the required drip rate in drops per minute.

Example

An infusion of 1 g of medication in 600 ml of glucose is to be given at a rate of 1 mg/min. Using an IV set delivering 20 drops per ml, calculate the rate in drops per minute to be used.

Method

    1 g is contained in 600 ml
    Therefore 1 mg is contained in 600 ml ÷ 1000 = 0.6 ml
       (because 1 g = 1000 mg)
    Amount required in one minute = 0.6 ml
    But 1 ml is given in 20 drops
    Therefore 0.6 ml is given in 0.6 × 20 drops = 12
    Drip rate = 12 drops per minute

6. An 800 ml infusion contains 1 g of a prescribed drug. Calculate the drip rate in drops per minute needed to deliver 1 mg/min. (Assume an IV set rate of 20 drops per ml.)

7. An X infusion contains 5 g in 500 ml Glucose. What drip rate in drops per minute would deliver 1 mg/min. (IV set gives 1 ml per 20 drops.)

8 You have an infusion of 4 g in 500 ml of solution. A dose of 2 mg/min is required. Find the drip rate in drops per minute, assuming an IV set delivering 20 drops per ml.

9 An infusion of strength 5 mg in 400 ml is to be used at a dosage of 4 microgrammes per minute. Calculate the drip rate in drops per minute if the IV set delivers 1 ml per 20 drops. (Take particular care with units.)

10 Find the drip rate in drops per minute to deliver 450 ml of medication over four hours. (IV set delivers 20 drops per ml.)

11 Using an IV set giving 15 drops per ml, find the drip rate required to deliver a 500 ml unit of plasma over five hours.

Example

An IV set delivers 20 drops per ml at a rate of 10 drips per minute. Find the time taken to deliver 200 ml.

Method

20 drops give 1 ml
1 drop gives 0.05 ml (1 ÷ 20)
10 drops give 10 × 0.05 = 0.5 ml (per min)
Minutes for 200 ml = 200 ÷ 0.5 = 400
Time taken = 400 mins = 6 hrs 40 mins

12 An IV set delivers 20 drops per ml. Find the time taken to deliver:
  a 150 ml, using a drip rate of 20 per minute.
  b 300 ml, using a drip rate of 30 per minute.
  c 500 ml, using a drip rate of 25 per minute.

— 14 —
Ratio

Both the sharing of money and the strength of solutions are often expressed in the form of a *ratio*.

Example

Because of differing needs, it had been decided to share a £10 000 grant between two charities X and Y in the ratio of 3 to 2, which is expressed as 3:2. This means that for every £3 received by X, the Y charity would receive £2.

Method

The £10 000 is divided into five equal parts (3 + 2) and then distributed as three parts to X and two parts to Y:

£10 000 ÷ 5 = £2000
X receives 3 × £2000 = £6000, Y receives 2 × £2000 = £4000

The separate parts when added together must equal the total to be shared: £6000 + £4000 = £10 000.

Instead of first dividing by 5 and then multiplying by 3, the sum can be set out as a fraction:

$\frac{3}{5}$ × £10 000 = £6000 and $\frac{2}{5}$ × £10 000 = £4000

N.B. Order is important. X was named first and must therefore receive the three parts which are also placed first in the ratio.

1 If an amount of money is to be shared in the ratio of 4:5, into how many equal parts should the money be first divided?
2 If the fraction method was preferred in working out question **1**, give the two fractions that would be used in multiplying the amount of money to be shared.
3 A council voted to give an extra £24 000 to social services, who decided to share the money between Residential Care and Day Care for people with physical disabilities in the ratio of 5:1. Calculate the amount of money each group would receive.
4 A hospital shared a £1 000 000 grant between three wards – Green, Blue and Red – in the ratio of 2:3:5, according to their needs. Find the amount each ward would receive.

Care must be taken not to confuse the ratio of amounts with the parts of the whole.

Example

A lotion is given as one part stock and six parts water.
a Express the proportion of stock to water as a ratio.
b Express the stock as a part or fraction of the lotion.

Method

    a Stock to water = 1:6
    b Altogether there are seven parts (1 stock, 6 water). Therefore there is one part stock in seven parts of lotion or 1 in 7 ($\frac{1}{7}$).

To summarise: In mixing a lotion where the ratio of stock to water is 1:6, there is one part of stock in seven parts of lotion.

5   Write the following ratios in the form 2 in ?, 1 in ?, etc.
    a 2:3   **b** 1:5   **c** 7:5   **d** 3:10

6   Write the following as ratios:
    a 2 in 3   **b** 3 in 8   **c** 2 in 7   **d** 4 in 9

7   500 ml of solution are to be produced from stock and distilled water in the ratio of 1:9. Find the amount of stock and the amount of water used.

8   The strength of a solution is given as 1 in 4.
    a Express the amount of stock as a fraction of the solution.
    b Calculate the amount of stock used in 1 litre of solution.

9   900 ml of distilled water were used in producing a solution of stock and water mixed at a ratio of 1:3.
    a How many ml of stock were used?
    b Find the total amount of solution produced, expressed in:
    (i) ml, (ii) litres.

10 If one part in eight of a lotion was stock, calculate how many litres of lotion were produced in using 250 ml of stock.

Simplifying ratios

Example

Two day centres were to share a sum of money in a ratio to the numbers of people they served. The number attending the Bradgate Centre was 44 and the number attending the Clifford Street Centre was 36. Express the number attending the Bradgate and Clifford Street Centres as a ratio, to enable the money to be fairly shared.

Method

The ratio of the Bradgate to Clifford Street Centres is 44:36.
The ratio 44:36 can be simplified to 11:9 by dividing both numbers by 4, as in cancelling.

11 Express the amounts £3000 to £4000 as a ratio in the lowest terms.

12 If 250 ml of stock were combined with 500 ml of distilled water, give the ratio of stock to water in simplified terms.

13 A council set aside £55 000 to be divided between Elwood Lodge (30 residents) and Breedon House (25 residents), both of which housed people with learning difficulties. If the division of funds was to be made according to the number of residents, calculate the amount each hostel would receive.

14 In trying to match resources to needs, a county fire officer gathered information that showed the number of call-outs his three main stations received over a given period. Titchmouth logged 375, Paulton logged 625 and Ainsbury logged 875. If the allocation was dependent on call-outs, express the logged numbers as a ratio.

— 15 —

Percentages – expressing one amount as a percentage of another

Our daily lives are dominated by percentages. Wage increases, cost of living changes, sales reductions, deaths from diseases, road accidents, smoking, health finance, treatment of patients, waiting lists – these are all measured in percentage terms. However, the sad fact is that most people do not understand them. Make certain you are not one of those people.

Parts of a whole can be expressed as a fraction, decimal or percentage, depending on what information is being presented. If a survey found $\frac{13}{40}$ of a county population was over 60 years of age, it is better expressed as 32.5% and more easily remembered.

Per cent means per hundred, therefore fractions must relate to 100, e.g. $\frac{1}{2} = \frac{1}{2}$ of 100 or 50%, $\frac{1}{4} = \frac{1}{4}$ of 100 or 25%.

Example a

Express £3 as a % of £4.

Method

As a fraction it would be shown as $\frac{3}{4}$
To express a fraction as a percentage multiply by 100:

$\quad \frac{3}{4} \times \frac{100}{1} = 75\%$

Example b

Express £2.40 as a % of £12.00.

Method

When any one amount contains pence, change both amounts to pence:

$$\frac{240}{1200} \times \frac{100}{1} = 20\%$$

Example c

Asking a similar question in a different way: What % of £1500 is £500?

Method

$$\frac{500}{1500} \times \frac{100}{1} = 33.33\% \text{ (corrected to 2 decimal places)}$$

Example d

What % of 2.5 litres is 375 ml?

Method

When any one amount contains ml, change both amounts to ml:

$$\frac{375}{2500} \times \frac{100}{1} = 15\%$$

1. Express £30 as a % of £200.
2. Express £1600 as a % of £6400.
3. Express 25p as a % of 75p.
4. Express 255 ml as a % of 500 ml.
5. Express £1.50 as a % of £6.00.
6. Express 500 ml as a % of 5 litres.
7. Express 200 g as a % of 4 kg (change to g).
8. Express 2.4 litres as a % of 5 litres (change to ml).
9. Express 300 mg as a % of 25 g (change to mg).
10. Express 2.5 kg as a % of 3.5 kg.
11. What % of £14 is £10?
12. What % of 3 litres is 2 litres?
13. What % of 6 kg is 300 g?
14. What % of 3 lbs is 4 oz?
15. What % of 1 g is 750 mg?
16. What % of 1 stone is 7 lbs?
17. What % of 1 kg is 1 lb? (If 1 kg = 2.2 lbs)

18 What % of 8 m is 200 mm?
19 What % of 1 yard is 2 feet?
20 What % of 2 mg is 500 microgrammes?

— 16 —
Percentages, fractions and ratios

In the previous chapter, it was shown that to express a fraction as a percentage the fraction was *multiplied* by 100. Therefore, to express a percentage as a fraction the percentage must be *divided* by 100.

Example a
Express 5% as a fraction.

Method
$$\frac{5}{100} = \frac{1}{20}$$

Example b
Express 3.5% as a fraction.

Method
$$\frac{3.5}{100} = \frac{35}{1000}$$ (by multiplying top and bottom lines by ten to remove the decimal point)
$$= \frac{7}{200}$$

Express the following % as fractions:

1 10%   2 35%   3 17%
4 18%   5 7.5%   6 2.25%

Sometimes we need to express a % strength of a lotion as a ratio.

Example
A solution was a mixture of stock and distilled water. If the strength of this solution was 5% (stock to solution), express the strength as a ratio.

Method

$5\% = \frac{5}{100} = \frac{1}{20} = 1$ in 20 (one part stock to 20 parts solution) or 1 : 19 (stock to water)

This helps to explain the meaning of the term 'percentage strength'. When one part stock is mixed with 19 parts distilled water, the stock represents $\frac{1}{20}$ of the mixture or 5%.

7 Express 4% in the form: **a** 1 in ? **b** 1 : ?
8 Express 8% in the form: **a** 2 in ? **b** 2 : ?
9 Express 12.5% in the form: **a** 1 in ? **b** 1 : ?
10 Express 7.5% in the form: **a** 3 in ? **b** 3 : ?

— 17 —
Solution strength

Sometimes a stock is already diluted before further mixing or dilution takes place.

For example, if a litre of solution was required at 50% dilution and the stock was already at 50% dilution, then clearly the stock could be used without further reduction in strength.

The following formula could be used to find the *volume* of stock required:

$$\text{Stock volume} = \frac{\text{solution strength}}{\text{stock strength}} \times \text{solution volume}$$

Using this formula for the example above:

$\frac{50\%}{50\%} \times 1$ litre = 1 litre of stock   (since the stock was already at the required strength)

If 1 litre of solution was required at strength 25% and the stock was strength 50% then, using the formula, the amount of stock needed would be:

$\frac{25\%}{50\%} \times 1$ litre = 0.5 litre = 500 ml

As you would expect, by adding an equal amount of distilled water, the strength of the stock has been halved.

1. If 1 litre of solution of strength 10% was required, what volume of stock at strength 50% would be needed?
2. Find the volume of a medication (strength 20%) required to produce 200 ml of solution at a strength of 3%.
3. If the final strength of 500 ml of solution was to be 0.2%, calculate the amount of stock at strength 0.8% required.
4. A stock of strength 5% is to be used in producing 2 litres of a stock and normal saline solution at a strength of 0.1%. Calculate:
   a the volume of stock required.
   b the volume of normal saline required.
5. Find **a** the stock, **b** the normal saline, required to produce 5 litres of a 0.4% strength solution with a stock of strength 10%.

When the strengths are given as ratios, then the formula used is the same but the setout is slightly different.

Example

500 ml of solution at strength 1 in 50 is required using a drug of strength 1 in 10 and normal saline. Find:
a the volume of drug required.
b the volume of saline required.

Method

$1 \text{ in } 50 = \frac{1}{50}$ $1 \text{ in } 10 = \frac{1}{10}$ (chapter 16)

$$\text{Volume of drug} = \frac{\text{solution strength}}{\text{drug strength}} \times \text{solution volume}$$

$$= \frac{1}{50} \div \frac{1}{10} \times 500$$

$$= \frac{1}{50} \times \frac{10}{1} \times \frac{500}{1}$$

$$= 100$$

a Volume of drug required = 100 ml
b Volume of normal saline required = 400 ml

Calculators make this work much easier, but first you must convert the fractions to decimals (chapter 12). The above example could then be entered as:

$0.02 \div 0.1 \times 500$ (where $0.02 = \frac{1}{50}$ and $0.1 = \frac{1}{10}$)

6. The strength of 1 litre of a stock and normal saline solution was 1 in 20. If the stock strength was 1 in 4, calculate the amount of:
   a stock, **b** normal saline.
7. A drug of strength 1 in 3 was used in producing 800 ml of an infusion (strength 1 in 60). How many ml of drug were used?

Sometimes 'strengths' are a combination of ratios and percentages, but the formula used remains the same.

Example

Two litres of solution are required at a strength of 1 in 200. The stock to be added to normal saline is of strength 40%. Find the volume of stock to be used.

Method

$$1 \text{ in } 200 = \frac{1}{200} \quad 40\% = \frac{40}{100}$$

$$\frac{\text{solution strength}}{\text{stock strength}} \times \text{solution volume} = \frac{1}{200} \times \frac{100}{40} \times \frac{2000}{1}$$

$$= 25 \text{ ml}$$

8   A medication of strength 60% is used to produce 600 ml of solution to a strength of 1 in 100. Calculate the volume of medication used.

9   If 1.5 litres of solution (strength 1 in 80) are produced by using stock (strength 25%) and normal saline, find:
    a the volume of stock used.
    b the volume of normal saline used expressed in: (i) ml, (ii) litres.

— 18 —
Percentage problems

1   If the strengths of two solutions are given as **a** 3 in 25, **b** 1 in 400, calculate each % strength.

2   The weekly costs in running a day nursery were as follows:

   Food £86, Rates £40, Power £25, Wages £850, Rent £160, Insurance £25, Telephone £25, Miscellaneous £39

   Calculate the wage cost as a % of the total costs.

3   By the year 2005 in the UK, it is estimated there will be an extra 500 000 people over 75 years of age compared to the 3.3 million in 1993. Show this increase as a % of the 1993 figure.

4   The number of cot deaths had remained fairly constant at 2 per 1000 live births until 1991 when the death rate dropped to 1.3 per 100 live births. Express as a % of live births: **a** the pre-1991

figure, **b** the 1991 figure. (*N.B.* since 1991 the death rate has dropped even more.)

5   Find the % increase in student doctors requested if there was a call for the intake to rise by 270 to 4470 per annum.

6   According to an official survey, the cases of tuberculosis in the UK had risen from 5204 in 1990 to 5861 in 1992. Give this increase as a % of the 1990 figure.

7   A London borough had estimated that for 1992/3 its total budget for Social Services would be £55 million, and of that figure £37 million was required for the adult client group.
    a Express the adult requirement as a % of the total budget.
    b Is there a need to work this sum in millions? (discuss)

8   The full cost of a dental examination in 1992 was £5.00, but an NHS patient was charged £3.75. Calculate the % reduction received by an NHS patient.

9   In 1993, Britain had 1.4 doctors per 1000 population, whereas Germany had 2.9. Express as a % the number of doctors per 1000 population in: **a** Britain, **b** Germany (try and give this answer by studying the answer to **a**).

10 The Knee High Playgroup had a maximum capacity of 18 children, but at Easter there were three vacancies. Express the take-up of places as a % of the allowed capacity.

11 Tom's salary as a trainee medical laboratory officer was increased from £6350 to £7000 per annum. Calculate the % increase he received.

12 **a** A kitchen assistant was required to mix 500 ml of cleaning fluid with 1.5 litres of water to achieve the correct concentration. Calculate the % strength of the mixed solution.
    b A concentration of 5 ml of stock was mixed with distilled water to produce 2 litres of solution. Calculate the % strength of the solution.

13 A man was diagnosed diabetic. Pre-diagnosis, his normal average diet of 2500 calories had a carbohydrate (CHO) content of 1100 calories. The man was counselled to achieve a CHO diet of between 50% and 55% of calorific intake.
    a Was his pre-diagnosis diet satisfactory?
    b He subsequently achieved a CHO content of 1350 calories in a total diet of 2500 calories. Express this CHO content as a % of his total intake.

14 In 1993, it was expected nationally that in 19 out of every 20 call-outs, vehicles would reach the scene of an emergency or incident within 19 minutes of the call being received. Show this expectation rate as a %.

15 A record maintained by a unit nursing officer showed that during January there were 750 hours of absence. Give the absence as a % of the total establishment of 18 400 hours.

— 19 —

Puzzles

Solving puzzles can be fun, surprisingly interesting and very useful in helping to understand arithmetical problems.

Example a

An allowance of £2000 is paid towards travelling expenses to a hostel housing 25 people. Assuming the same degree of funding, how much should be paid to a hostel caring for 38 people?

Method

Always reduce the number sharing to one:

If   25 receives 2000

Then   1 receives $\frac{2000}{25}$ (do not work out at this stage)

Therefore 38 should receive $\frac{38}{1} \times \frac{2000}{25} = £3040$

Example b

In one year a team of specialists dealt with 287 day patients. If this represented 35% of the total number of patients treated, how many patients were treated altogether?

Method

Total number of patients = 100%

If   35% = 287

Then   1% = $\frac{287}{35}$

Therefore 100% = $\frac{100}{1} \times \frac{287}{35}$ = 820 patients

Example c

600 mg of penicillin is supplied in 4 ml of stock solution. If a baby is prescribed 125 mg, how much stock solution should be 'drawn'?

Method

There are 600 mg in 4 ml

$$1 \text{ mg in } \frac{4}{600} \text{ ml}$$

$$125 \text{ mg in } \frac{125}{1} \times \frac{4}{600} \text{ ml} = 0.83 \text{ ml}$$

When using a calculator, the order of entries for **c** above would be:

$$4 \div 600 \times 125$$

In questions **2**, **3** and **4** set out the fractions only, do not work the sum out. Question **1** shows the answer required.

1. If 42 represents 3% of a group of people, find 1%.
   Answer $\frac{42}{3}$

2. If £119 represents 7% of an amount of money, find 1%.

3. If 200 mg of a drug are contained in 5 ml of solution, how many ml will contain 1 mg?

4. If 24 less accidents represents a 6% reduction, find 1%.

5. If £119 represents 7% of an amount of money, find: **a** 16%, **b** 100%.

6. If 200 mg of a drug are contained in 5 ml of solution, how many ml would contain: **a** 100 mg, **b** 175 mg.

7. If 24 less accidents over a month represented a drop of 6%, calculate the number of accidents there were in the previous month.

8. A 2 ml solution contains 600 mg of penicillin. Find how many ml should be 'drawn' to give a dosage of: **a** 100 mg, **b** 125 mg, **c** 180 mg.

9. It had been estimated that 15% of a liquid had evaporated. If 340 ml remained, calculate the original amount of liquid before evaporation, expressed in: **a** ml, **b** litres.

10. A hospital chef worked on a food cost of 40% of the selling price when pricing meals for the staff restaurant. What should be charged for a meal with a food cost of £1.20? (The selling price is 100%.)

11. Karen had reckoned on ordering seven bottles of orange juice per

week for the 28 toddlers attending her playgroup. After Christmas the number increased by eight children. How many bottles should Karen now order?

12 The costs involved in running a home for 30 residents were £1 200 000. Estimate the costs of running a similar home for 22 residents.

— 20 —
More about percentages – multiplying by a percentage

So far we have expressed one amount as a percentage of another, but sometimes we need to be able to find a given percentage of an amount. For example, we may need to know 3% of £120.

Remember, 1% is one hundredth and to find one hundredth we must divide by 100.

There are two methods of working this type of sum. Both involve dividing by 100 to find 1% and then multiplying to find the required percentage.

Example

Find 3% of £120.

Method A

$$3\% = \frac{3}{100} \text{ (chapter 16)}$$

$$\frac{3}{100} \text{ of } 120 = \frac{3}{100} \times \frac{120}{1} = \frac{18}{5} = £3.60$$

Method B

1% of £120 = £1.20 (quick division by 100)
Therefore 3% = £3.60 (multiply by 3)

1	1% of £300	**2**	1% of 800 ml
3	1% of £475	**4**	1% of 5 kg (change to g)
5	1% of £26	**6**	1% of 405 m

7   2% of £850
8   3% of 600 g
9   7% of £1250
10 5% of 15 litres (change to ml)
11 4% of £1.00
12 20% of 8 kg
13 30% of £2300
14 15% of 200 ml
15 35% of 400 mg
16 18% of 1 km (change to m)
17 2.5% of 200 litres
18 1.25% of £1000
19 9% of 1.5 kg (change to g)
20 7.5% of 2.8 litres (change to ml)

— 21 —

More percentage problems

1. Find the amount of stock used in producing 2 litres of solution with a strength of: **a** 1%, **b** 4%.

2. An electricity company announced a price increase of 6% on the previous year. Assuming no increase in consumption from one year to the next, calculate the estimated increase on last year's electricity bill of £950.

3. A local authority budgeted for a spending of £2 200 000 on Mental Health, of which 45% was set aside for salaries. Calculate the amount allowed for salaries.

4. A medical secretary earning £9000 per annum was awarded a 5.5% increase in salary. How much extra would she earn during the next year?

5. Residents at a home serving people with learning difficulties shared the first 60% of the meals and travel costs of any accompanying staff when making an outside visit. Two residents were accompanied by a member of staff when making a visit which involved costs per person of: travel £3.00, food £2.50. Find:
   a the money contributed by each resident towards sharing the costs of the member of staff.
   b the total money paid by each resident.

6. In a group of 25 Health and Social Care students, 40% intended to follow a career in nursing. How many indicated they would prefer other aspects of care?

7   Proteins represented 12% of an active young man's daily intake of 3000 calories. Find the calorific value of the proteins he consumed in 24 hours.

8   Find the payment an NHS dental patient would make for having a tooth crowned if she was charged 80% of the £86.00 cost.

9   Mr Smithers considered investing £250 000 in a nursing home to give him an estimated 7% return (profit) on his capital. However, present investments of this capital brought in £18 200 per year. On the known facts, which is the better use of his money and by how much per year? (discuss)

10 If 55% of the receipts from meals were made available to a hospital chef to spend on food and drink for the restaurant, calculate how much the chef will receive from sales of £2500.

— 22 —
Profit and loss

Although it may seem inappropriate to use terms such as profit and loss in the Health and Social Care sector, the growth of nursing homes, childcare and private treatment indicates that many organisations and individuals are investing in this area. Even in non-profit-making hospitals and services there may be 'profit centres' – areas identified where profits are sought to enable facilities to be provided elsewhere. For instance, a hospital might identify visitors' car-parking provision as a profit-making opportunity.

Net profit is the difference between the total costs and the total income received in providing a service.

Of course, if the costs exceed the income then the result is a loss.

A 'break-even point' is reached when the income equals the expenditure.

Example

The total costs involved in running a playgroup amounted to £5050. Over the same period of time, the total money received for providing this service was £12 000. Calculate the profit.

Method

$$\begin{array}{ll} \text{Total income} & 12000 \\ \text{Total costs} & 5045 \\ \hline \text{Net profit} & 6955 \end{array}$$

It is usual to express the profit as a % of the money received:

$$\frac{6955}{12000} \times \frac{100}{1} = 57.96\%$$

By expressing the answer as a %, comparisons can easily be made with other profit areas.

1. During one week, the takings from a hospital car park amounted to £2100. The cost of employing parking-ticket staff for one week was £520. Calculate:
   a the profit for the week.
   b the profit as a % of the takings.

2. The income and expenditure for the Glenmore Nursing Home during one year were as follows: total receipts – £286000, total expenses – £196000. Find:
   a the net profit as a % of the receipts.
   b the net profit as a % of the £750000 capital employed in setting up the home.

3. A manager was required to achieve a 6% profit on the running costs of his hospital in a year when they amounted to £100 200 000.
   a Write in words the costs in the year.
   b Calculate the expected profit.

4. A local council gave a grant of £287 000 for the provision of facilities for the hard of hearing, but at the end of the year it was found that expenses of £295 000 had accumulated. Find the difference in the two amounts and give the financial terms that should be applied to this situation.

5. The staffing costs of a day nursery amounted to £856 and all other costs amounted to £914. Calculate the total receipts over the same period if it was found that a break-even position had been achieved.

— 23 —
Costing a meal

Catering provides an important service within the Health and Social Care industry, but the method of accounting for this facility varies according to the area in which it operates. The scale of provision ranges from a playgroup snack to a full hospital catering service, including a restaurant for staff and visitors. Costs can be subsidized by a grant, or covered fully by a charge or allowance. Sometimes a profit is required. Whatever method is used, one factor is common – the costs need to be known and therefore need to be calculated.

A simplified costing sheet is reproduced below showing the costs in producing an Irish Stew.

COSTING SHEET

Dish Date..........

Irish Stew

	Quantity	Unit cost	Total cost
Meat			
Stewing lamb	2.55 kg	£2.42/kg	£6.17
Poultry, fish			
Greengrocery			
Potatoes	2.4 kg	£0.10/kg	£0.24
Celery	1 head	£0.65	£0.65
Button onions	600 g	£3.90/kg	£2.34
Onions	600 g	£3.20/kg	£1.92
Leeks	600 g	£1.50/kg	£0.90
Dairy			
Dry stores			
Total			£12.22
Number of portions			24
Cost per portion			£0.51

Use the prices given below to work out the following questions.

      beer   £1.00 per litre

butter	£2.52 per kg
button mushrooms	£3.90 per kg
caster sugar	£0.95 per kg
cheese	£4.40 per kg
eggs	£0.80 per dozen
flour	£0.36 per kg
haddock	£6.30 per kg
jam	£1.60 per kg
leeks	£1.50 per kg
margarine	£0.64 per kg
macaroni	£0.60 per kg
milk	£0.44 per litre
onions	£0.52 per kg
potatoes	£0.10 per kg
stewing beef	£4.48 per kg
stewing lamb	£2.52 per kg

Draw costing sheets and use them to work out the cost per portion of the following (work to the nearest penny):

1 *Carbonnade of Beef*
(8 portions)

800 g beef
30 g caster sugar
400 g sliced onion
500 ml beer

2 *Macaroni Cheese*
(10 portions)

875 g macaroni
200 g butter
150 g flour
2.8 litres milk
875 g cheese

3 *Haddock in Cheese*
(20 portions)

3.5 kg haddock
750 ml milk
125 g margarine
125 g flour
875 g cheese

4 *Queen of Puddings*
(24 portions)

3 litres milk
600 g caster sugar
150 g butter
300 g jam
18 eggs

If a profit is required, it is usual to express that profit as a percentage of the selling price.

Example

A dish cost £0.80 to produce. If the dish was sold for £1.20, calculate the profit as a % of the selling price.

Method

Selling price	£1.20
Cost	0.80
Profit	0.40

$$\frac{40}{120} \times \frac{100}{1} = \frac{100}{3} = 33.33\%$$

Profit is 33.33% of the selling price.

This profit is known as the *gross profit*, because only the cost of materials (food) has been deducted from the sales.

5. A dish which cost £0.75 to produce is sold for £1.00. Calculate: **a** the profit, **b** the profit as a % of the selling price.

6. Find the profit as a % of the selling price if a dish which cost £1.80 to produce is subsequently sold for £2.00.

7. The total ingredient cost of a dish was £2.30. If the dish was priced at £5.75 in the staff restaurant, calculate the profit as a % of the selling price.

Making a profit

Where a chef is required to make a given % profit, the calculation is as follows.

Example

A chef is required to achieve a profit of 10% on the selling price of a dish costing 45p. Calculate the selling price.

Method

$$\text{Cost} + \text{Profit} = \text{Selling price } (100\%)$$
$$45p + 10\% = 100\%$$
Therefore $\quad 45p = 90\% \, (100\% - 10\%)$

Therefore $\quad 1\% = \dfrac{45}{90}$ (chapter 19)

Therefore $\quad 100\% = \dfrac{100}{1} \times \dfrac{45}{90} = 50$

The selling price should be 50p.

8. Calculate the selling price of a dish costing 60p if the chef is required to make a profit of 20% on the selling price.

9. Find the price that should be charged for a dish costing £1.26 to achieve a profit of 25% on the selling price.

10. The kitchen in a hospital was looked upon as a profit centre when supplying food and drink to visitors and staff. Calculate the charge that should be made for a cup of coffee in order to make a profit of 80% on the selling price, if the ingredient cost was 12p.

11. To help cover the staff costs in cooking a meal at a day nursery, it was decided to aim for a profit of 50% on the charge for a lunch. How much should be charged for a meal costing £0.85 to achieve the intended profit margin?

12. Find the charge that should be made for a portion of Queen of Puddings (question **4**) in order to realise a profit of 60% on the selling price.

— 24 —
Preparing a budget

Weather forecasting and financial budgeting have a lot in common – a little bit of luck, a great deal of guesswork, backed up with facts, figures and experience, all combined to try and predict the future.

1. Look back over the past year and list the differences between your *expectations* in terms of income and expenditure, and your *actual* income and expenditure. Try and account for the differences (e.g. increase/decrease in income, transport costs, meals, council tax, gas and electricity).

A considered approach is far better than a 'wait and see' attitude which may have ruinous consequences. It is essential to plan ahead, but do not forget the value of past records and experience of trends.

2. Produce your own budget for the next 12 months based on money you expect to receive (e.g. earnings, grant, pocket money, gifts) and your projected outgoings (e.g. accommodation, meals, travel, clothes, entertainment, holidays, subscriptions). Balance your income and expenditure each month and carry the balance forward to the next month. Try and estimate increases and decreases based on experience.

3. The estimated income and expenditure in March for the Care Free Day Nursery was as follows:

    Fees charged for children attending – £5200, Receipts from meals – £1580, Wages – £2560, Rent – £1200, Power – £200, Repairs – £300, Insurance – £250, Rates – £200, Food and drink – £900, Miscellaneous – £250

    a Copy the proforma overleaf and enter the above details in the March column to find the net profit.

	March	April
Income		
Fees		
Meals		
Total (X)		
Expenditure		
Wages		
Rent		
Power		
Repairs		
Insurance		
Rates		
Food and drink		
Miscellaneous		
Total (Y)		
Net profit (X − Y)		

  b The estimates for April, based on the March figures were:

Fees − £520 increase, Receipts from meals − £50 increase, Wages − one extra member of staff (£520), Rent − 5% increase, Power − 2% decrease, Repairs − £50 decrease, Insurance − 10% increase, Rates − 4% increase, Food and drink − 5% increase, Miscellaneous − no change.

Complete the proforma to find the forecasted profit for April.

  c If there was a possibility of losing five children in July, which headings under income and expenditure could be directly affected? (Ignore miscellaneous. Staff ratio to children is strictly enforced.)

4 At 1 January, the estimated income of the Resteasy Private Nursing Home for the coming year was £350 000, and the expenditure was expected to be £300 000. After six months the interest to be repaid on a loan of £100 000 was unexpectedly reduced from 18% per annum to 16%, and it became clear that other annual costs would be 1.5% higher than the original estimation.

  a (i) Find the original expected profit.
    (ii) Express this profit as a % of the £500 000 capital invested in the home.
  b (i) Find the actual profit.
    (ii) Express this profit as a % of the capital.

— 25 —
Stock control

There is a wide range of goods carried in the store rooms of the Health and Care Industry, and a good system of stock control should be implemented. Readily available stock is essential — and could be a matter of life or death.

Stocks may vary from towels and linen to dangerous drugs, and from food and drink to life preserving blood. The storekeeping may also differ according to the value and security involved, but the principles of good stock control are common.

A good storekeeper will implement some or all of the following:

a Goods received should always be checked and agreed with the delivery note, and any discrepancies or damage should immediately be taken up with the supplier or lessor.

b A good system of recording stock levels should be implemented, whether manual or by the increasingly popular computer. The diagram below shows a simple stock card or bin card.

Item ....   **No.** ....

Maximum Stock 20   **Minimum Stock** 5

Date	Receipts	Issues	Balance	Remarks
6 Feb	10	2	8	
8	10	–	18	
12	–	7	11	
15	5	9	7	

N.B. In this example the stock level must not fall below five or rise above 20. (Computer stock records can be programmed to indicate when high or low stock levels have been reached.) On 15 February there are seven items in stock. In the Remarks column notes can be made, e.g. price, supplier's name.

c Stock rotation should ensure that old stock is used first, with due regard to the deterioration rate or obsolescence of certain items.

d Suppliers' estimation of supply dates should be taken into consideration for re-ordering purposes. Use past records to establish a minimum stock level for each item.

e Irregular but constant stocktaking should be practised by management to check that the theoretical stock (on paper) agrees with the physical stock (look and see).

f Maintain a good storage environment. This usually requires different areas for perishables and non-perishables.

g An efficient stock-issuing system will help to prevent pilfering, but not hinder the dispensing of stock.

h Finally, it is worth remembering that stock is an asset with a monetary as well as a beneficial value, and as such it should be protected by a reliable system of security.

1 Design and complete a bin card using the following information:

    Item – towels (leased)
    10 January – 40 delivered
    12 January – 15 issued
    14 January – 16 issued
    17 January – 30 delivered
    19 January – 18 issued
    21 January – 15 issued

Check the final balance with the answer at the back of the book.

2 Complete a bin card from the following:

    Item – beans (tins, large)
    Maximum stock   20   Minimum stock   3
    3 September – 12 received
    3 September – 3 issued
    6 September – 5 issued
    10 September – 12 received
    12 September – 4 issued

3 Using the information provided in question **2**, calculate the value of the stock in hand on 12 September, given that the cost of a tin of beans was £0.75.

4 If the towels in question **1** were leased at £0.28 each, find the amount to be invoiced (charged) for the period 10 January–21 January.

5 Give reasons for having a maximum stock level. (discuss)

6 Give reasons for having a minimum stock level. (discuss)

— 26 —
Graphs

Information presented in graphical form is both simple to produce and easy to understand. Graphs have been used in hospitals for many years to record patients' temperature, pulse rate and blood pressure, often on the same graph. Income and expenditure can be plotted to give quick comparisons but, better still, trends can be easily detected.

Study the graph below which shows a patient's temperature from 5 April to 8 April inclusive.

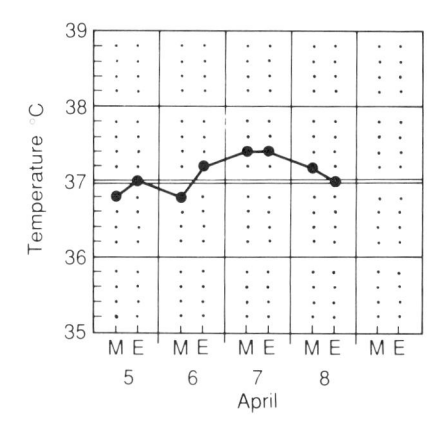

The vertical axis represents the temperature in degrees centigrade. As there are five graduations per degree, each graduation represents 0.2°C.

The horizontal axis refers to the days when temperatures were taken, M = morning, E = evening.

Normal temperature is represented by the double lines at 37°C.

Example

Find the temperature on the morning of 6 April.

Method

Find the plot vertically above M on 6 April. Trace the position horizontally to the left of the plot to find the temperature 36.8°C.

1. What was the recorded temperature on the evening of 7 April?
2. When was a normal temperature recorded?
3. What was the lowest temperature recorded?
4. What was the highest temperature recorded?

Notice how much easier it is for a doctor or nurse to check on the movement of a patient's temperature by studying a graph rather than reading figures.

The following diagram shows the finances associated with a Drop-In Centre over a period of ten weeks, and illustrates how it is possible to present two sets of figures on the same graph.

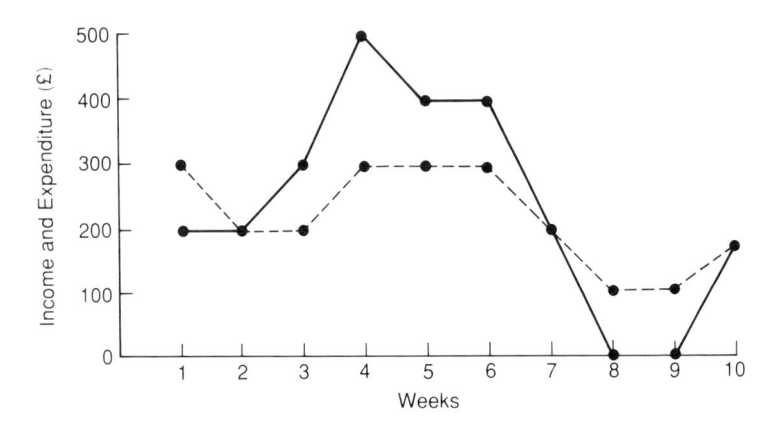

The continuous line shows the income received, while the broken line depicts the expenditure over the same period of time.

5. Find the difference between income and expenditure during the fourth week.
6. In which weeks was a break-even point reached? (Remember, this is when income equals expenditure.)
7. When would you assume the centre was closed?
8. What was the difference in income and expenditure over the ten weeks, and say whether it resulted in a profit or loss.
9. The data below gives the pulse rate of a patient over four days. Present the data in the form of a graph using the following specifications: let the vertical axis show the pulse rate with a range of 60–90; let the horizontal axis show the days sub-divided into morning and evening.

Days	8 Nov		9 Nov		10 Nov		11 Nov	
	M	E	M	E	M	E	M	E
Pulse	70	72	70	76	72	74	80	74

10 Plot the attendance at the Tot Up Day Nursery from the following details: Monday – 20, Tuesday – 23, Wednesday – 24, Thursday – 27, Friday – 23.

Straight line graphs

Straight line graphs can be used to produce conversion tables or any information where values are constant. Study the graph below which plots kilogrammes against the equivalent pounds (assuming 1 kg = 2.2 lbs).

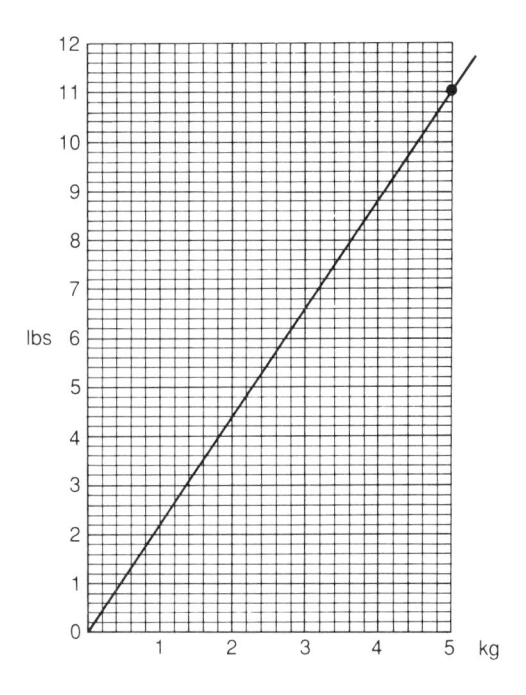

A plot is made at zero kg against zero lbs.

A second plot is made at 5 kg against 11 lbs (5 × 2.2 – any value of kg could be used but 5 kg is simple to show in lbs).

A straight line joins the two plots.

Example

Find the equivalent value of 3 kgs in lbs.

Method

Find the position vertically above 3 kgs where the line intersects, and read off the value in lbs horizontal with that position. Answer 6.6 lbs.

11 Find the equivalent value in lbs of 2 kgs.
12 Find the approximate value in kgs of 4 lbs.
13 Express 4 kgs in lbs.

The larger the graph, the more accurate the answer.

14 Produce a straight line graph to show equivalent centigrade and fahrenheit values. Plot from 0°C to boiling point.

Bar charts (block graphs or histograms)

Although the line graph is better for illustrating ongoing (changing) information, static information is best depicted by a *bar chart*.

The chart below shows the sharing out of funds by a local authority.

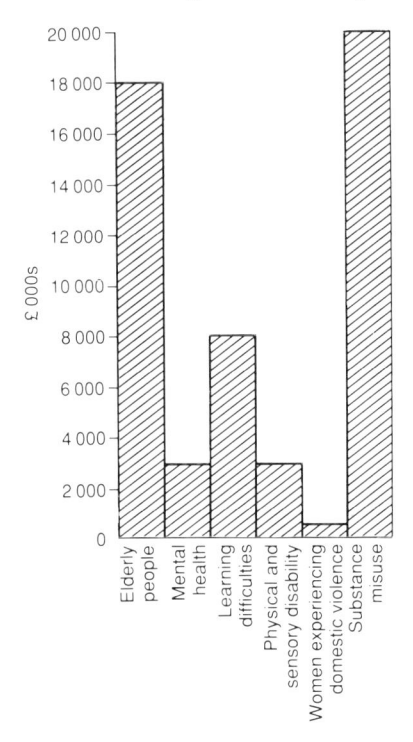

At the side of the vertical axis £000s indicates that the figures in that column represent thousands of pounds, i.e. they should be multiplied by 1000. For example, the figure 16 000 represents £16 000 000.

Note how quickly a comparison of the fund allocation can be made.

15 How much money was allocated to people with learning difficulties?

16 Which group received the biggest funding?

17 Which group do you think received a small grant although the local authority is not directly responsible for its funding?

18 A local authority provided residential homes for the elderly. Over one week the occupancy levels were as follows:

Home	white	black	red	green	blue
Places	50	60	25	70	64
Occupied	49	39	23	56	48

Draw a bar chart to show 'at a glance' the *percentage* occupancy from the above information.

19 A college kept a record of the jobs obtained by students on Care courses over the last three years. Present this information in the form of a bar chart.

Nursing – 62, Child day centres – 30, Care administration – 10, Au pairing – 12, Residential homes – 25, Hospital assistants – 15, Other – 18

— 27 —

Pie charts

A clear and convenient way of showing the division of a whole into parts is by the use of a *pie chart*. This is particularly useful when the allocation of resources needs to be illustrated.

The diagrams below show a circle (360°) representing the whole of the money set aside for Mental Health by a district county council, and a pie chart showing the actual division of that money.

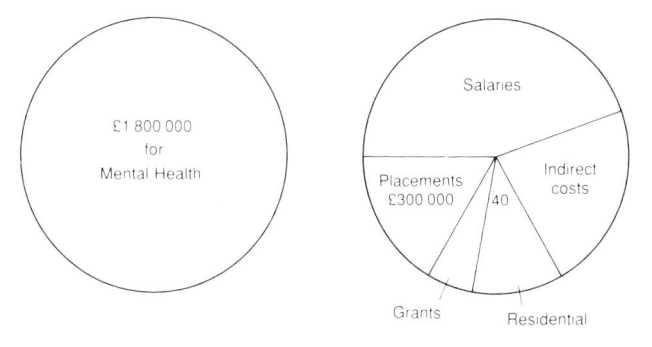

As 360° represents £1 800 000, then 1° represents
£1 800 000 ÷ 360 = £5000.

By the use of a protractor, the Residential segment can be measured as 40° or £200 000 (40 × £5000). The money allowed for Placements is represented by 60° or £300 000.

1. Calculate the value of the Grants segment if it is measured as 20°.
2. Find the degrees of the arc of the segment showing Salaries, if it represents a value of £800 000. Check your answer with a protractor.
3. How many degrees represent the Indirect Costs and what is the value?

A pie chart can be made more effective by differently colouring the individual segments.

4. Draw a pie chart to show the 'percentage of energy received from nutrients' in a particular diet, given that protein provided 10%, fat provided 40% and carbohydrates provided 50%. (Find 10% of 360° for protein etc.)
5. A day centre found that the total expenditure in a year was as follows:

    Wages – £52 000, Accommodation – £12 000, Food – £5000, Repairs and renewals – £2000, Miscellaneous – £1000

    Draw a pie chart to illustrate the expenditure and say what the size of the segment is for Accommodation. (Find the total expenditure to find the value of one degree.)
6. A hospital caterer reckoned that for every £1 he raised in income he spent 60p on food and drink, 25p on wages, 10p as a charge to accommodation/power, and the difference went to reserve funds. Produce a pie chart to represent the division of the £1 of income, and state the size of the segment depicting the wages.

— 28 —
Centigrade and fahrenheit

Despite a determined attempt to remove fahrenheit from British temperature measurement, weather reports still include both centigrade (C) and fahrenheit (F), and most older members of the population are more at ease with temperatures expressed in fahrenheit.

A sizeable proportion of the British public still think of the normal blood temperature as 98.6°F and they consider a weather forecast of 70°F as a prediction for a reasonable day. Carers must respond to this dual measurement, and be able to express temperatures in both centigrade and fahrenheit for the foreseeable future.

$$0°C = 32°F \text{ (freezing point)}, 100°C = 212°F \text{ (boiling point)}$$

For students who prefer formulae, the following will prove useful:

$$°F = \tfrac{9}{5}C + 32 \text{ (multiply °C by } \tfrac{9}{5} \text{ and add 32)}$$

$$°C = (F - 32)\tfrac{5}{9} \text{ (subtract 32 from °F and multiply by } \tfrac{5}{9}\text{)}$$

1 Express the following in °F:
   a 10°C **b** 45°C **c** 37°C

2 Given that $\tfrac{9}{5} = 1.8$, express the following in °F:
   a 20°C **b** 0°C **c** 38°C

3 Express the following in °C:
   a 59°F **b** 212°F **c** 95°F

A simple method is to use a graph as shown in the diagram below.

Let the vertical axis represent the fahrenheit values from 32 to 212 and let the horizontal axis represent the centigrade values from 0 to 100.

Using two values we know – freezing point and boiling point – we plot them to produce a straight line graph.

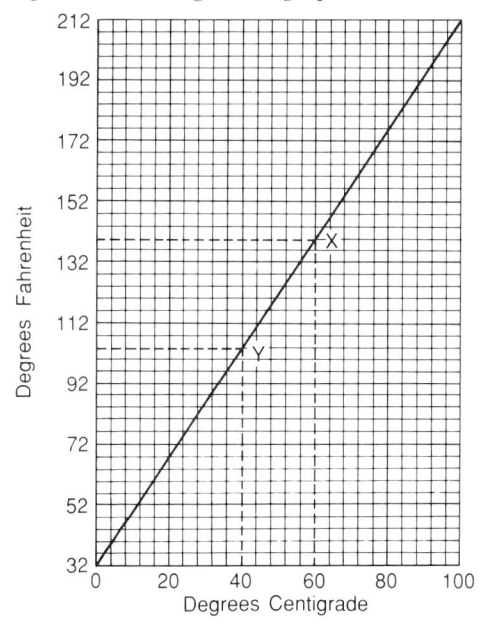

— 59 —

Example

Convert 60°C to °F.

Method

Plot the point vertically above 60°C on the line at X. Find the position horizontally level with X and read the answer – 140°F.

Example
Convert 104°F to °C.

Method

Plot the point horizontally from 104°F on the line at Y. Find the position vertically below Y and read the answer – 40°C.

Use your graph to work out the following:

4 Convert to degrees fahrenheit: **a** 90°C, **b** 35°C, **c** 45°C.

5 Convert to degrees centigrade: **a** 41°F, **b** 176°F, **c** 77°F.

– 29 –
Computer spreadsheets

Computers make calculating easy, provided you know what information to put into the machine in the first place. In this respect they are very similar to calculators. A calculator will not tell you how to solve a problem, but it will give an accurate answer every time using the information that is entered (the input).

A computer or calculator cannot make a decision about when to add, subtract, multiply or divide. Take percentage problems for instance. In some cases an amount is multiplied by 100 (express £3 as a % of £15), at other times an amount is divided by 100 (find 8% of 4 kg). Unless the basic arithmetical method is known, it is pointless using a calculator or computer.

Hardware and *Software* are terminology used to describe the essential make-up of a computer package. Hardware refers to the main computer (memory and processing units, display screen, keyboard, printer, the electronic workings). Software refers to the programs that can be entered into the computer, either semi-permanently by way of a hard disk, or temporarily using a floppy disk (similar to playing tapes on a cassette player). These enable the user to carry out certain functions.

The *spreadsheet* is one particular piece of software that is the subject of this chapter.

Computers work by using programs, and writing these programs is a skilful job. However, a spreadsheet is a pre-written program that can be 'loaded' into a computer and then acts as a simple and useful tool for accessing and calculating certain data.

To use spreadsheets to best advantage, confidence in basic arithmetic as set down in this book should be achieved.

Let us imagine that income and expenditure are known and the balance is required. The simple arithmetical formula needed is:

Income − Expenditure = Balance

A typical spreadsheet might show the following columns:

Column A	Column B	Column C
		(A − B)
Income	Expenditure	Balance

N.B. The user types in the headings Income, Expenditure, Balance and enters the formula (A − B).

Provided the user has entered the correct formula and method for that spreadsheet in column C, the answer will always be correct.

Extra columns can be added if required. Column D could show the balance as a % of income, i.e. C ÷ A × 100. You can have fun creating your own formula depending on your needs.

The rows of columns can be added to give total income, expenditure or balance. The spreadsheet can be used to calculate horizontally or vertically.

By selecting from a 'menu' displayed at the head of the page, it is also possible to produce data in the form of a graph or pie chart which can then be preserved by means of a print-out.

Spreadsheets can be set up for the user by someone else so that the user merely enters (in this case) the income and expenditure, and the balance will be immediately displayed. However, there are times when a user needs specific information and it is simple and self-satisfying to be able to input a formula.

Programable calculators work on exactly the same principle − basic arithmetical knowledge is important to use these exciting tools.

1. Using the headings A, B, C, in the previous example, write a formula in column D to give the expenditure as a % of income.
2. Let column A represent the number of ambulance call-outs. Column B represents the number of incidents reached in the time allowed.
   a Write a formula for column C to show the number of incidents not reached on time.
   b Write a formula for column D to show the % of call-outs reaching the incidents in the time allowed.
3. Let column A give the total staff hours available and column B the hours of staff absence. Write a formula for column C to give % absenteeism.
4. If column A represented the number of people surveyed and column B the number of 'non-smokers', write a formula for column C to show % of 'smokers'.

More columns can be added if required. In question **4** above, a column could be added giving the national % of people with 'smoking' related diseases for comparison. Another column could be added showing related death rates.

Here is a further use of formulae in a spreadsheet, showing the availability and percentage occupancy of beds in three wards of Gray's Hospital.

Gray's Hospital

Tuesday, Week 40

	A	B	C	D	E
1	Ward	Beds	Patients	Spare beds	% Occupancy
2	Kildare	B2	C2	B2 − C2	C2 ÷ B2 × 100
3	Finley	B3	C3	B3 − C3	C3 ÷ B3 × 100
4	Jekyll	B4	C4	B4 − C4	C4 ÷ B4 × 100
5	Totals	B2 + B3 + B4	C2 + C3 + C4	D2 + D3 + D4	C5 ÷ B5 × 100

Rows

N.B. Only the shaded area is displayed on screen.

Notice how row 1 is used to indicate the contents of the columns A, B, C, D and E.

B2 refers to the position (cell) of column B and row 2 etc.

Row 1 is entered and stored in the memory.

Column A is entered and stored in the memory.

The formulae in columns D and E and row 5 are stored in the memory.

The user now enters the number of beds and patients in columns B and C, leaving the computer to calculate and display the answers as below.

Gray's Hospital

Tuesday. Week 40

	A	B	C	D	E
1	Ward	Beds	Patients	Spare Beds	% Occupancy
2	Kildare	20	18	2	90.00%
3	Finley	30	25	5	83.33%
4	Jekyll	15	12	3	80.00%
5	Totals	65	55	10	84.62%

Spreadsheets are excellent for presenting and comparing data clearly and simply.

— 30 —
Mixed tests

Test 1

1 **a** $3.245 + 117.8 + 10.07$ **b** $14.08 \text{ kg} - 12.275 \text{ kg}$

2 Cancel: **a** $\frac{54}{81}$ **b** $\frac{52}{78}$

3 **a** Pam was paid £4.25 per hour working as a phlebotomist in her local hospital. Find her gross wages in a week in which she worked the following hours:

   Monday   9.00 a.m. – 12.45 p.m.
   Tuesday   9.00 a.m. – 11.30 a.m.   1.45 p.m. – 4.00 p.m.

Thursday 9.00 a.m. – 12.45 p.m.
Friday 1.45 p.m. – 4.00 p.m.

  b Explain the difference between 'gross wages' and 'net wages'.

4 The diagram below shows the dimensions of a room. Find the floor area of the room.

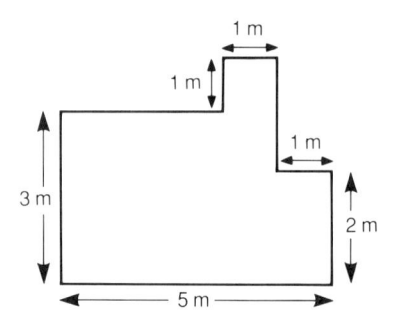

5 Using the tables on page 9 express: **a** 13.05 litres in ml, **b** 83 oz in lbs and oz.

6 Find the arithmetical mean of 4, 9, 17, 9, 8, 25.

7 How many days inclusive was the stay in hospital of a patient who was admitted on 28 September and discharged on 4 October.

8 The petty cash float of the Tots Playgroup was maintained at £20. Calculate the cash that must be given to the petty cash bookkeeper after a week in which the costs of purchases were as follows: £5.00, £3.25, £0.85, £2.66, £7.05.

9 An intravenous set delivers 1 ml every 20 drops. How much is delivered per drop?

10 Money is shared between two hostels as follows: Exbury – £50 000, Beaulieu – £85 000. Express these amounts as a ratio of Exbury to Beaulieu in the lowest possible terms.

Test 2

1 **a** $2\frac{3}{8} \times 7\frac{1}{5}$ **b** 0.5×0.5

2 Write $\frac{3}{25}$: **a** as a decimal fraction, **b** as a percentage.

3 Using the tables on page 9, express 14 kg in lbs.

4 Share £5550 between three groups in the ratio 3 : 2 : 1.

5 How many ml of distilled water should be added to 225 ml of stock to produce 3.8 litres of solution?

6 If 35 kg represented 14% of an amount, find 100%.

7 Mr Williams was an in-patient at the Gables. On 1 July he had a balance of £25.45 in his personal cash account. During July he spent the following amounts:

5 July – Chocolates £2.60, 6 July – Travel £1.75, entrance to show £5.25, 12 July – Chocolates £1.45, 16 July – Trip £3.00, 19 July – Pub lunch £4.20, 23 July – Sweets £1.92, 25 July – Book £2.40

Produce a cash account for Mr Williams showing the balance as at 31 July.

8 What is the meaning of the term 'break-even point'?

9 Using the ingredient costs on page 47, calculate the total costs of 4 dozen eggs, 1.5 kg margarine, 5 kg potatoes, 3 kg stewing beef, 3 litres milk.

10 A Fluid Balance Chart showed that over 24 hours Mr Howe had a fluid intake of 2.5 litres and an output of 2700 ml.
   a Would the balance be described as 'positive' (more input than output), or 'negative' (more output than input)?
   b Give the fluid difference in: (i) litres, (ii) ml.

Test 3

1 Find 85% of: **a** £145, **b** 12.5 kg.

2 Express 9.7382 corrected to 2 decimal places.

3 Sarah, who was employed as a dental receptionist, was awarded a 5% pay rise which brought her an extra £405.40 per annum. Calculate Sarah's new annual salary.

4 By studying the records of 19 patients, it was found that their nights stayed in hospital were as follows:

   3, 7, 2, 12, 1, 7, 3, 8, 6, 10, 5, 5, 7, 4, 2, 10, 12, 7, 3

   Find the median number of nights stayed in hospital.

5 A particular lotion is produced by adding stock to distilled water in the ratio of 2:5. Calculate the amount of stock needed to be mixed with 850 ml of water.

6 Of the 34 children attending a day nursery, 14 had a midday meal. Express the number of children who did not have a midday meal as a % of the total attending.

7 A meal cost £2.50 to produce. Find the price to be set by the chef if he needs to achieve a profit of 60% on the selling price.

8 In times of rising prices, would it be more sensible to increase your stock holding levels or decrease them? (discuss)

9 **a** Is 28°F above or below freezing point?

b Would 28°F be expressed as a plus or a minus on the centigrade scale?
10 A fire brigade found that in 95% of cases an appliance reached the scene of a fire within the time expectation. Calculate the number of calls received over a given period if on 30 occasions the time limit was exceeded.

Test 4

1 Express 12.75 litres: **a** in pints, **b** to the nearest gallon.
2 **a** $16\frac{1}{4} \div 8\frac{2}{3}$ **b** $13.08 \div 0.12$
3 The pie chart shows the distribution of funds between four homes for the elderly. If the total money available was £1 000 000, find the amounts allocated to each home.

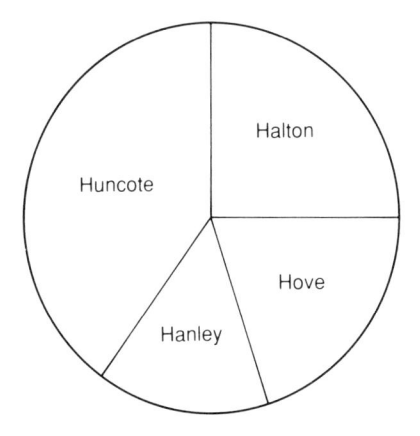

4 Calculate the amount of stock at a strength of 50% needed to produce 1 litre of solution at a strength of 20%.
5 In the course of her work, a health visitor used her car to travel 350 miles. If she was allowed 33p per mile for official duties, calculate the total claim for mileage allowance she should submit.
6 After carrying out a survey, a medical centre found that 73 out of 225 patients questioned were 'smokers', although 39 had only stopped smoking during the last 12 months. What percentage of those questioned were 'non-smokers'?
7 Complete the following sentence: '14:28 is the same as 1 part in ... parts.'
8 The Brychester Health Authority was presented with a cheque by the local Lions Club for £29 750, which was to go towards the

purchase of an ambulance. If the value of the cheque represented 85% of the cost, find the total amount required to buy the new ambulance.

9 Draw a bin card to include the following details:

Item – Sheets
Maximum stock – 50
Minimum stock – 15
1 February – Received 42, issued 10
2 February – Issued 5
3 February – Issued 12
5 February – Received 30, issued 9

Find the balance at the end of the day on 5 February.

10 The income of the Redgrove Nursing Home over one month amounted to £36 000. Costs over the same period were: Wages £8250, Overheads £18 350.
   a Calculate the excess of income over expenditure and express this as a percentage of the income.
   b Give an appropriate term for the excess.

Test 5

1 13.25 litres − 9.476 litres. Write the answer in: **a** litres, **b** ml.

2 Find 2.5% of 14.25 kg and give the answer to the nearest gramme.

3 Express: **a** 15°C in °F, **b** 50°F in °C.

4 Construct a line graph showing a patient's temperature changes from the following information:

23 April (a.m.) 36.8°C, (p.m.) 37.2°C
24 April (a.m.) 36.8°C, (p.m.) 37°C
25 April (a.m.) 37°C, (p.m.) 37.6°C
26 April (a.m.) 37.4°C, (p.m.) 37.2°C

5 A nursing home charged £205 per week caring for an elderly resident, but the local authority agreed to pay the first £185 of this amount. Calculate the cost borne by the local authority as a percentage of the full charge.

6 A survey of British households in 1991 revealed that single-parent families rose from 8% in 1971 to 19% in 1991. What was the numerical increase per 1000 households over the 20 years?

7 Pauline worked the following hours as a medical centre receptionist:

| Monday | 8.30 a.m. – 11.15 a.m. | 1.15 p.m. – 4.00 p.m. |
| Tuesday | 8.30 a.m. – 11.15 a.m. | 2.00 p.m. – 4.00 p.m. |

  Wednesday 11.15 a.m. – 2.00 p.m.
  Thursday 9.00 a.m. – 12 noon 2.15 p.m. – 4.30 p.m.
  Friday 8.30 a.m. – 11.15 a.m. 1.15 p.m. – 3.30 p.m.

Calculate Pauline's weekly gross wage if she is paid £4.16 per hour.

8 A doctor prescribed a drug for a child at a dosage rate of 30 mg for each kg of weight. Find the dose the child should receive if she weighed 19.5 kg.

9 **a** If $3\frac{1}{4}$ feet are equivalent to 1 metre, how many square feet are contained in 1 square metre?
 b A local authority ruled that a day nursery must have a minimum floor space of 70 square metres. If the plan below shows the dimensions of the nursery, use the answer to **a** above to say whether permission would be given to use the facility.

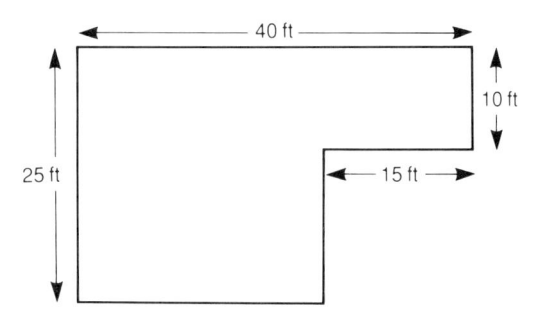

10 The weights of six babies born on one day were as follows:

 4 kg, 3 kg, 2.98 kg, 3.45 kg, 4.1 kg, 3.5 kg

 a Find the mean average weight.
 b Express the average weight corrected to the nearest lb.

Test 6

1 Cancel: **a** $\frac{54}{321}$ **b** $\frac{143}{187}$

2 Express $\frac{17}{20}$ as a decimal fraction.

3 The Shirley Self-Help Group maintained an imprest petty cash account with a float of £20.00. On 1 May the balance brought down was £4.19 and the transactions during the week were as follows:

 1 May – Received cash to make up float
 2 May – Purchased folder £0.75, staples £1.15
 4 May – Travel £2.10

5 May – Postage stamps £3.24
6 May – Photocopying £2.90

Produce a petty cash account, enter the above transactions and balance the account on 6 May to find the amount that should be carried down on 8 May.

4 A 500 ml transfusion of blood is to be given over 3 hours. Assuming an IV set rate of 15 drops per ml, find the required drip rate in drops per minute.

5 An au pair was offered a year's employment in France at a rate of 2000 francs per annum (plus accommodation and meals). After 3 months, her employer increased the payment she received by 3%. Calculate the total amount the au pair received over the 12 months.

6 A district nurse was sent to an 'Updating Conference'. She received a subsistence allowance (meals etc.) of £5.60 per day and a travel allowance of 18.6p per mile or second class train fare (whichever was the cheaper). The course commenced on 27 July and ended on 4 August. The return fare was £22.75 and it was 72 miles from her home to the conference centre. Find the total claim the nurse was able to submit.

7 A grant of £14 350 was shared between three centres in the ratio of the numbers registered as able to attend. If 56 were registered at Greensleaves, 112 at Knighton and 224 at Rowlett, calculate the money each centre received.

8 The *essential* formula to calculate the size of the fuse (in amps) needed on an electric plug is as follows:

$$\text{Amps} = \text{Wattage} \div \text{Volts or} \left(A = \frac{W}{V} \right)$$

   a Find the amps flowing through a plug using an electrical appliance rated at 1000 watts (1 kilowatt) on the UK 240 volt system.
   b Which of the following common fuses should be used in the plug mentioned in **a**: (i) 3 amp, (ii) 5 amp, (iii) 13 amp?

N.B. Never use a fuse with a rating higher than necessary. (discuss)

9 If a supplier was reliable and could deliver stock when required, should that encourage a storekeeper to maintain a high or low level of stock? Say what advantages there are in putting your decision into operation.

10 A meal was required at 19.20 hours. If the cooking time was estimated at two and a quarter hours and a further 10 minutes were allowed for serving, calculate the time the chef should

commence the cooking: **a** using the 24 hour clock, **b** using the 12 hour clock.

Test 7

1. Express: **a** 14.54 m in mm, **b** 50 g in kg.
2. What % of £8.50 is £3.40?
3. The following bar chart shows the numbers of reported road accidents in a county over a 6 month period.

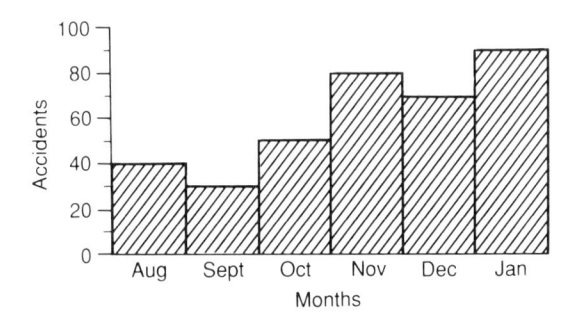

   a Find the number of road accidents in September.
   b Give a possible reason for the increase in October.
   c Calculate the mean average of accidents per month.
   d Does the graph suggest there was an increase in accidents due to wintry weather conditions?

4. The 36-room Bramwell Nursing Home charged residents £315 per week. During the week commencing 4 September, 25 rooms were occupied. Calculate:
   a the percentage occupancy in the week commencing 4 September.
   b the weekly income during that week.
   c the maximum possible income in a week.
   d the estimated annual staff costs if they were reckoned as 25% of the maximum possible income.

5. A receptionist at a surgery with three doctors made appointments for each doctor at the rate of two patients each quarter hour. The appointment hours were 8.30 a.m. – 10.15 a.m. on Mondays, Tuesdays and Fridays; 8.30 a.m. – 10.45 a.m. on Wednesdays; 4.15 p.m. – 6 p.m. on Thursdays. How many patients could be seen by appointment in one week?

6. If the standard prescription of insulin for a diabetic child is 2 units per kg of the child's weight and 1 ml contains 100 units:
   a calculate the number of units required for a child weighing 28 kg.

b express as a decimal the part of 1 ml of insulin used.

7 To assist in gauging the size of overalls to order for stock purposes, a housekeeper decided to use the modal average of overall sizes of the present staff, which were as follows:

14, 12, 18, 16, 16, 14, 20, 10, 14, 18, 10, 12, 14, 14, 16, 14, 12, 16

What was the modal size?

8 In 1993, the Department of Transport reported that in the previous year deaths in road accidents fell by 7% to 4229 and there was also a fall in serious injuries from 51 836 to 49 244. Calculate:
a the deaths from road accidents pre-1992 (to nearest whole number).
b the % drop in serious injuries.

9 A chef produced 20 portions of Smoked Haddock Pie using the following ingredients:

2.25 kg haddock
3.25 kg potatoes
125 g flour
1.5 litres milk
750 g cheese
2.5 kg leeks

Using the costs on page 47 prepare a costing sheet to find:
a the cost per portion (to the nearest penny).
b the price to be charged if the chef was required to achieve a 55% profit on the selling price.
c If the price was eventually set at £2.50 for simplicity, find the profit per portion as a % of the selling price.

10 Looking ahead to next year, Francine estimated the numbers of children attending the Newton Church Playgroup would average 22 full-time equivalents, each bringing in £15 per week. Charges for meals would result in a profit of £25 every week. This year's staff cost of £9450 was likely to increase by 2% and the present overheads of £4500 were reckoned to rise by 5%. Based on the estimates made by Francine, produce a budget for the next 12 months and say what the financial position is expected to be, assuming the playgroup is open for 50 weeks in the year.

Test 8

1 **a** Write in words – 2 300 057.
 b Write in figures – nine hundred and two thousand, two hundred.

2 **a** Increase 80 by 17%. **b** Decrease 950 kg by 12%.

3 Mrs Jones took a doctor's prescription for iron tablets to her local chemist who advised her that, although the NHS prescription would cost £4.75, the tablets could be bought without the need for a prescription for £3.09. As Mrs Jones took advantage of the chemist's advice, what was her percentage saving on the NHS price?

4 Two litres of solution (stock and normal saline) of strength 3% are required.
   a What volume of stock of strength 50% is needed?
   b What volume of normal saline is needed?

5 A Health and Fitness report on the relationship between height and weight in women stated that for a woman of height 5 ft 6 in the average weight is 9 st 2 lbs, with an acceptable range of 8 st 2 lb to 10 st 6 lbs.
   a Express: (i) the height given above in metres, (ii) the average weight in kg.
   b According to the report, is it acceptable for a female of height 5 ft 6 in to weigh 63 kg?

6 The Social Services Department of Harchester gave the following information in the annual report:

   Number of alarms installed   628
   Calls Received:
   A   Genuine emergencies   54
   B   Assistance required   72
   C   Inappropriate requests   26
   D   Technical faults   12
   E   False alarms   280

   a Express as a percentage of the total calls: (i) genuine emergencies, (ii) false alarms.
   b Complete the following sentence: 'Technical faults occurred in 1 in ... calls.'
   c In the list A to E above, why do they not total to the number of alarms installed?

7 A typical heading in national statistics is £m or £000000. This indicates that the value of whole numbers in that column is in millions of pounds. Study the figures below which show a selection of Special Transitional Grants for Community Care in the year 1993–4.

	£m
City of London	0.053
Cornwall	6.649
Lancashire	20.081
Leicestershire	8.764
Kent	19.198

| Isle of Wight | 2.314 |
| Suffolk | 6.135 |

Give these amounts in pounds, e.g. the Grant for Leicestershire is £8 764 000.

8 A social worker was helping Mr and Mrs Wilson to determine if they were entitled to any Income Support. As they were both between 68 and 75 years of age, they qualified for a 'couples' extra pension premium of £29.00 per week to add to their £69.00 pension allowance. From the total of these allowances and premiums must be deducted payments earned – the difference is the Income Support they would receive. The social worker found Mr Wilson received a pension of £4721.60 per year. Find the Income Support, if any, the Wilsons should receive each week.

9 Mary Redgrave was a resident in a home for disadvantaged persons. On 1 June her personal cash account showed she had a balance of £5.06. During June she had the following transactions which were recorded by the staff:

  4 June – Bought sweets £2.10
  6 June – Received £12.00
  10 June – Bus fares £3.25, meal £3.60
  18 June – Purchased comb £0.75
  23 June – Outing £2.40, chocolates £0.85, postcard £0.15, stamp £0.24
  28 June – Purchased pen £1.04

Complete Mary's cash account for June, showing the balance to be carried down on 30 June.

10 A regional Family Health Service Authority was to use a spreadsheet to show the percentage of girls under the age of 15 who had not received the Rubella injection as a precaution against catching German Measles. The spreadsheet would give the figures for each doctor's practice in the region. Column A showed the number of girls under 15 years of age. Column B showed the number of girls under 15 years of age who had received the Rubella injection.
  a Using the details above, write a formula for column C which would provide the information required.
  b Can you think of other columns that should be added to give important information on this subject?

Test 9

1 **a** $8.892 \div 0.26$ **b** 45.7×1.8

2 How many 35 ml bottles can be filled from 2.5 litres?

3 A 500 ml infusion contains 2 g of a prescribed drug. Calculate the drip rate in drops per minute needed to deliver 1 mg/min (using an IV rate of 20 drops/ml).

4   A medication was proportionately related to weight. If a man of weight 70 kg required 28 ml of a medication, find the amount required by a man of weight 75 kg.

5   List reasons against carrying a large stock of one particular item.

6   A course of antibiotics consisted of 56 tablets, one to be taken four times each day. If the course commenced on the morning of 28 September, on what date should the last tablet be taken?

7   **a** Produce a pie chart to illustrate the division of the total annual costs shown below in running a 'Drop-In' centre within a London Borough.

	£
Staffing	35000
Food	3000
Transport	1000
Activities	2375
Stationery, telephone, postage	1000
Volunteers expenses	625
Miscellaneous	2000

   b What is the size of the arc representing 'Activities'?

8   A decision was taken to carpet the waiting area at the Elton Community Health Centre. A plan of the floor is shown below.

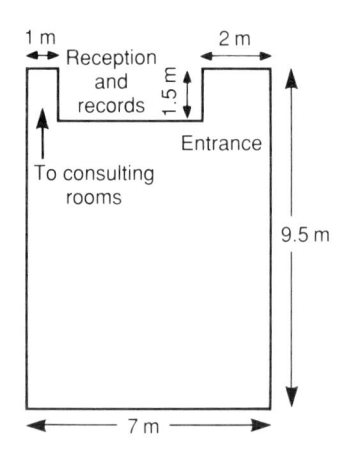

It was recommended that rolls of 1 metre wide, hard-wearing carpet costing £17.50 per square metre should be used.
   a Assuming no allowance is required for edging and it is not practical to use off-cuts to fill in odd areas, find the most

economical method of laying the carpet and calculate the total cost.

b The Heath Centre later negotiated an 8% discount off the original price. Find the price finally agreed.

9 A diabetic diet sheet gives the following information:

1 g protein converts to 4 calories
1 g fat converts to 9 calories
1 g CHO converts to 4 calories

a Calculate the total calories in a diet in which the 60 g of protein made up 10% of the calorific content.
b If the protein in **a** represented the intake of a diabetic man whose CHO daily intake should be equal to 55% of the diet, find the recommended weight of CHO he should consume.

10 Statistics produced by a college showed that over the last three years the ratio of male to female students on Health and Care courses remained constant at 2:5. Enrolments revealed that 28 male students had enrolled on the new course for September. Based on past records, how many female students could be expected?

Test 10

1 From 1.65 litres take 325 ml and give the answer in: **a** litres, **b** ml.

2 How many microgrammes are equal to: **a** 1.75 grammes, **b** 0.25 milligrammes?

3 Figures released in 1993 showed that over 12 months the number of patients waiting for hospital treatment between one and two years dropped by 25% to 64 875, but for patients waiting up to one year the numbers increased by 12% to 952 000. Calculate the numbers waiting in 1992: **a** between one and two years, **b** less than one year.

4 A chef found that 85 out of 221 patients requested vegetarian meals.
a Express the vegetarians as a fraction of the total patient population in the lowest terms possible.
b Express the answer to **a** as a percentage.

5 All calls at an ambulance control room were logged using the 24 hour clock system. During the early evening, John looked at his watch at seventeen minutes to seven to note a call.
a At what time should John log the call?
b If John signed off at 22.30 after a 6 hour shift, write the time he signed on using the 12 hour clock.

6 On three days per week, Margaret used her car to deliver 'meals on wheels'. She delivered to six households and averaged nine miles each day. If Margaret received a vehicle allowance of 18.4 pence per mile, calculate the money she claimed for using her car during February 1994.

7 In presenting a fire brigade charter, the county fire service stated that, in the event of an emergency, householders living in small towns in the area could expect a fire engine within 10 minutes of a call being received. During one week, the times taken to arrive at the scenes of emergencies were as follows:

3 mins – 4 calls, 4 mins – 8 calls, 5 mins – 16 calls, 6 mins – 6 calls, 7 mins – 7 calls, 8 mins – 20 calls, 9 mins – 15 calls, 10 mins – 12 calls, 11 mins – 10 calls, 12 mins – 5 calls, 15 mins – 1 call.

   a Find the mean average time taken.
   b What percentage of calls exceeded the expected time?

8 Charges for the Richmond Private Hospital were as follows:

Standard room £200 per night
Executive room £220 per night
Guest room £80 per night
Early admission   Before 8 a.m.   £145
                  8 a.m. – 12 noon £18 per hour
Late discharge   After 6 p.m.   £145
                  2 p.m. – 6 p.m.   £18 per hour

Mrs Edwards was admitted to the Richmond Hospital at 4.15 p.m. on 27 January and her husband spent the first night in the guest room. Mrs Edwards was discharged at 4 p.m. on 3 February. Calculate the total room charges if Mrs Edwards occupied an executive room.

9 Sandra worked as a medical secretary for 18 hours per week and received a *pro rata* salary (proportion of full-time salary according to hours worked). The full-time equivalent post was £10 500 per annum for a 35-hour week. Graham worked 18 hours per week as an administrative assistant at a medical centre, for which he was paid £3.85 per hour for a 48-week year.
   a Calculate the annual remuneration each received.
   b If Sandra was allowed 3 weeks paid leave each year, find who received the better rate per hour and by how much.

10 In a solution of a drug and normal saline, the proportion of the drug to the solution was 3 parts in 5. Find:
   a the percentage strength of the solution.
   b the ratio of stock to saline.

Test 11

1 Write: **a** 14.2346 corrected to 2 decimal places, **b** £2.2382 to the nearest penny.

2 Find 7% of 13.25 kg and express the answer in grammes to the nearest gramme.

3 In sanctioning a day nursery, a local authority laid down the following criteria relating to staffing levels:

Age Group	Ratio (staff/children)
Under 2 years	1 : 3
2 to 5 years	1 : 6

In addition to the above, there must be a supervisor.

Rates of pay	
General staff	£3.20 per hour
Supervisor	£4.25 per hour

a Find the total minimum staff costs for a morning session (9 a.m. – 12 noon), when the following children were in attendance:

Age Group	Attendance
Under 2 years	14
2 to 5 years	12

b (i) Which age group could take more children without incurring extra staff costs?
 (ii) How many more children could be taken at no extra cost?

4 Mrs Cook and Mr Brown were to become residents at the Tree Tops Nursing Home where the fees were £215 per week. After making an allowance of £13 each week for personal expenses, the local authority agreed to pay the fees, less any pension received and less certain income from savings. Every £250, or part of £250, in savings above £3000 would result in the local authority support being reduced by £1. Mrs Cook and Mr Brown both had weekly pensions of £60, but Mrs Cook had savings of £4900.

a If the local authority made a contribution of £168 per week to the fees for Mr Brown, calculate the contribution made by the local authority to Mrs Cook's fees.
b How much interest would be earned per year on capital of £250 in a deposit account paying interest at the rate of: (i) 5%, (ii) 8%.
c What rate of interest would earn £52 per year on a deposit of £250?
d Mr Brown had savings of £3000. On the information presented, which of the two residents received the most advantage from their savings? (discuss)

5 It is quite common to value stock in hand at the lower of cost price (the price at which the goods were bought) and market price (today's price). Use this method to value:

—77—

a seventeen articles in stock that were originally purchased one month ago at £2.56 each, but would now cost £2.95 each.
  b nine articles that were purchased last week for £7.28 each, but have today been reduced to £7.05 each.

6 In the UK, over-the-counter pharmacy sales of health supplements (multi-vitamins, cod liver and fish oils, garlic etc.) grew from £201 000 000 in 1991 to £227 130 000 in 1992.
  a Calculate the % rise on the 1991 figure.
  b Find the amount spent per head in the UK on Homoeopathic medicine each year, if it is estimated we spend 4% of the £5.42 paid by the average Dutch person. (Answer to nearest penny.)

7 During one week a fire sub-station received the following number of calls between 6 p.m. and 6 a.m. each night:

  4, 5, 4, 4, 20, 3, 4

  a Find the mean average number of calls per night.
  b Find the median average number of calls per night.
  c Which of the two answers above is more relevant for statistical purposes?

8 A cleaning fluid used in a nursing home was a mixture of 350 ml of disinfectant and 3 litres of water. Calculate the percentage strength of the fluid.

9 A social services committee decided to share this year's allowance of £1 800 000 for hostels in the ratio of last year's expenditure which was: Swithland £600 000, Charnwood £500 000, Cropston £400 000. Find the amount each hostel would receive.

10 The treasurer of the LittleUns playgroup was preparing the accounts for the year. Incoming fees from parents brought in £5000 and expenditure amounted to £5750. Later, when the financial details of the meals provision was taken into account, the treasurer proclaimed that a break-even point had been achieved. If the expenditure on food had been £250, what were the receipts for the meals?

Test 12

1. **a** $4\frac{3}{4} \times 2\frac{6}{7}$ **b** $5\frac{2}{3} \div 6\frac{4}{5}$

2. Nowadays, the commonly accepted value for a billion is one thousand million. Write in figures the estimated National Health Service prescription drugs bill for 1994 of £3.33 bn.

3. **a** Draw a line graph for use as a length conversion indicator in a maternity ward, to show the equivalent values of millimetres and inches. Let the plots range from 305 mm (12 inches) to 610 mm.
   b Read off the number of inches that are equal to 405 mm.
   c Read off the number of mm that are equal to 21 inches.

4. According to research, the private sector provided two-thirds of the 231 000 nursing home beds in 1992, and that was worth £2.2 bn to the private operator. Calculate to the nearest £1000 the value of a private bed.

5. Stock ampoules contain 40 mg of medication in 2 ml of solution. What volume of solution is required to administer 25 mg?

6. Michael Thomas, the Head Chef of the Highfield NHS Hospital, was allowed £1.67 per day to cater for each of the 550 patients that on average were resident in the hospital on any one day.
   a Calculate the total catering allowance received by the chef in a normal year.
   b On talking to a chef colleague, he found that his catering allowance amounted to 28% of the money received per patient in the nearby private hospital. What was the daily patient allowance in the private hospital? (Answer to the nearest penny.)

7. The Devonshire Hostel for Disadvantaged Persons maintained a petty cash account for day-to-day expenditure of amounts less than £5.00. The float was set at £20.00 and balanced each week. On Saturday 3 August, the balance carried down was £12.78. Enter the following transactions and balance the account on 10 August.

    5 August – Balance brought down
    5 August – Received cash to make up float
    6 August – Travel £1.80
    7 August – Postage £2.25, batteries £2.62
    8 August – Flowers £3.70
    9 August – Travel £1.45, printer tape £4.75
    10 August – Lubricant £1.05

8. The committee of the Newtown Day Nursery were planning ahead to next year. They had before them a proforma set out as overleaf, showing this year's estimates.

	Estimate
Grant	1 000
Fees income	36 000
Meals income	2 500
Total income	
Staff costs	24 250
Administration	2 800
Power	1 250
Rent/rates	5 000
Food/drink	2 000
Telephone	400
Repairs	1 000
Miscellaneous	2 000
Total expenditure	
Excess income/expenditure	

In fact, this year's estimates had differed from the actuality as follows: income was better than forecast by £1250, causing an increase in staff costs by £250; repairs were less than anticipated by £100, but miscellaneous expenses were greater by £250.

a Find the excess of income over expenditure in the original forecast.

b Extend the proforma to include a second column and enter the actual income and costs to find the excess of income over expenditure.

Based on the actual figures, the committee budgeted for next year as follows:

Grant – no change
Fees – £38 000 (if charges remain unchanged)
Meal receipts – increase of 10%
Staff costs – increase of 5% to implement new ratios/rates
Administration – no change
Power – increase of 8%
Repairs – £850
Rent/rates – increase of 5%
Food and drink costs – increase of 6%
Telephone – no change
Miscellaneous – increase of 10%

    c Extend the proforma to a third column and enter the budgeted income and costs to find excess of income over expenditure (if any).

The committee considered the excess was too small to allow for unforeseen circumstances and decided to find a further £2000. By tighter controls in administration and telephone costs they hoped to save £50 in each of these areas, and reluctantly they agreed to raise the fees to make up the difference.

    d By what percentage must the fees be increased?
    e Extend the proforma to a fourth column and enter the final proposed budget, including the excess of income over expenditure.

9 Figures produced by the National Congenital Rubella Surveillance Programme in 1993 showed that reported cases of German Measles in pregnant women had risen from seven occurrences in 1992 to 18 confirmed by blood tests for the first six months in 1993. A survey by the council also showed that an immunisation target of 90% of schoolgirls had been achieved by only 39.4% of district health authorities.

    a If the trend of German Measles continued throughout 1993, find the percentage increase in cases for the whole of that year over the 1992 figure.
    b What percentage of district health authorities failed to meet the immunisation target?
    c How many in every 10 000 schoolgirls were expected to be immunised?

10 In competing local hospitals, the weights of seven babies born within five minutes of the start of the new year were as follows:

    2.9 kg, 3.1 kg, 2.9 kg, 3 kg, 2.7 kg, 3.3 kg, 3.2 kg

Find the weight representing the following averages:

(a) mean   (b) median   (c) mode.

Answers

1 Addition (page 1)

1	770	**2**	1110	**3**	853	**4**	4659
5	2453	**6**	6492	**7**	5900	**8**	12643
9	16 hrs 53 mins	**10**	10 hrs	**11**	25 hrs 29 mins	**12**	8 hrs 33 mins
13	£0.09	**14**	£0.28	**15**	0.001 kg	**16**	0.025 kg
17	0.856 kg	**18**	0.005 litres	**19**	0.755 litres	**20**	0.038 m
21	0.485 m	**22**	**a** 12.235 kg **b** 7.030 kg or 7.03 kg				
23	**a** 4.050 litres or 4.05 litres **b** 6.005 litres						
24	£193.55	**25**	£344.03	**26**	4.623 kg	**27**	10.699 kg
28	22.931 litres	**29**	23.285 litres	**30**	562.639 m	**31**	1.508 m
32	47.418	**33**	22.415	**34**	244.056	**35**	69.14

2 Subtraction (page 3)

1	321	**2**	4325	**3**	4705	**4**	960
5	£21.26	**6**	£41.68	**7**	£189.08	**8**	£440.26
9	1.868 kg	**10**	2.899 kg	**11**	0.075 litres	**12**	0.875 litres
13	77.05 m	**14**	28.912 m	**15**	33 mins	**16**	1 hr 34 mins
17	1868 g	**18**	875 ml	**19**	28912 mm	**20**	£63.45
21	£90.63	**22**	1.45 kg	**23**	6.864 kg	**24**	0.595 litres
25	0.05 litres	**26**	70.249 m	**27**	0.844 m	**28**	£46.99
29	0.745 litres	**30**	8.939 m	**31**	0.17 kg	**32**	£6.67

3 Multiplication (page 4)

1	24	**2**	35	**3**	24	**4**	36
5	24	**6**	36	**7**	44	**8**	21
9	42	**10**	72	**11**	84	**12**	49
13	121	**14**	108	**15**	27	**16**	54
17	144	**18**	63	**19**	60	**20**	132
21	273	**22**	1080	**23**	2754	**24**	4970
25	18048	**26**	67680	**27**	73	**28**	16.38
29	14.68	**30**	£52.22	**31**	£150.72	**32**	£0.075
33	25 kg	**34**	2.4912 kg	**35**	5.112 litres	**36**	0.322 m
37	58.5 m	**38**	0.3575 litres	**39**	£532.59	**40**	525.6 m
41	252 litres	**42**	£123.40	**43**	235.5 kg	**44**	3711 litres
45	1275 m	**46**	7500 kg	**47**	168 litres	**48**	120 m
49	£10600	**50**	12.34 litres				

4 Division (page 5)

1	6	**2**	9	**3**	11	**4**	10
5	9	**6**	7	**7**	3	**8**	12
9	3	**10**	7	**11**	8	**12**	8
13	6	**14**	6	**15**	7	**16**	11
17	8	**18**	5	**19**	11	**20**	8
21	17	**22**	121	**23**	335	**24**	£845
25	268 kg	**26**	628 m	**27**	45	**28**	132
29	412	**30**	305	**31**	507	**32**	£8201
33	24.6	**34**	4.25	**35**	12.3	**36**	2.3
37	6.3	**38**	31.1	**39**	3.66	**40**	7.65
41	4.5625	**42**	3.13	**43**	2.318	**44**	20.56
45	2.24	**46**	145.444	**47**	17.1	**48**	£203.90
49	30.82	**50**	1.0253	**51**	£0.42	**52**	56.55 kg
53	0.08255 m	**54**	1.4 litres				

5 Metric and Imperial measurement (page 8)

1. 66 lbs
2. 13.888 litres
3. 3 lbs 2 oz
4. 5 tons
5. 161 km
6. 50 kg
7. a 3.5 mg b 0.0035 g
8. a 0.046 g b 46 000 microgrammes
9. a 1000 mg b 1 000 000 microgrammes
10. 1 000 000 mg
11. 3 tablets
12. $2\frac{1}{2}$ tablets
13. $\frac{1}{2}$ tablet

6 Problems using the four rules (page 10)

1. £147.52
2. 12 mm
3. £24.84
4. 630 mg
5. a 7.755 lbs b $7\frac{3}{4}$ lbs
6. £12.73 overspend
7. $2\frac{1}{2}$ tablets
8. a 4575 ml b 4.575 litres
9. a Private nursing home b Private nursing by £0.51 per hour
10. £177.10
11. £0.75
12. £5.75
13. 180 g

7 Area (page 12)

1. 108 sq m
2. 300 sq ft
3. 163.5 sq m
4. a No b 8 ft
5. £163.80
6. 4 sq m
7. 25 sq m
8. £200
9. 90 sq ft

8 Averages (page 15)

1. a £10 800 b 7 days
2. a 72.055 kg b 75.25 kg
3. £1 022 000
4. a mean b 8 mins
5. 4 fillings
6. a 9.75 mins b discuss, consider exceptional 40 mins c median time

9 Days inclusive (page 17)

1. 26
2. 14
3. 5
4. 47
5. 19
6. 8 June

10 Twenty-four hour clock (page 19)

1. 03.00
2. 13.00
3. 09.45
4. 18.15
5. 23.59
6. 00.05
7. 12.00
8. 21.24
9. 11 a.m.
10. 2 p.m.
11. 9.35 p.m.
12. 12.20 a.m.
13. 5.25 a.m.
14. 12.10 p.m.
15. 10 p.m.
16. 1 a.m.
17. 19 mins
18. 2 hrs 16 mins
19. 4.05 p.m.
20. 7 hrs 45 mins
21. 2 hrs 15 mins
22. 13 mins

11 Accounting for cash (page 20)

1. a £746.50 b £746.50
2. £13.50
3. £53.63
4. balance £3.00, cash £22.00
5. £2.77

12 Using fractions (page 24)

1. $\frac{4}{5}$
2. $\frac{3}{4}$
3. $\frac{4}{7}$
4. $\frac{6}{7}$
5. $\frac{1}{4}$
6. $\frac{1}{10}$
7. $\frac{1}{3}$
8. $\frac{1}{3}$
9. $\frac{1}{3}$
10. $\frac{2}{3}$
11. $\frac{1}{4}$
12. a $\frac{1}{60}$ b $\frac{1}{61}$
13. a $\frac{15}{80}$ b $\frac{3}{16}$ c $\frac{3}{19}$
14. $\frac{5}{2}$
15. $\frac{19}{5}$
16. $\frac{15}{4}$
17. $\frac{67}{9}$
18. $\frac{87}{8}$
19. $\frac{83}{12}$
20. $\frac{44}{25}$
21. $\frac{121}{3}$
22. $3\frac{1}{7}$
23. $3\frac{1}{5}$
24. $13\frac{2}{3}$
25. $9\frac{1}{12}$
26. $4\frac{3}{8}$
27. $8\frac{1}{2}$
28. $5\frac{1}{2}$
29. $8\frac{1}{2}$
30. $3\frac{3}{4}$
31. $\frac{3}{4} \times \frac{4}{5}$

32	$\frac{2}{7} \times \frac{5}{11}$	33	$\frac{15}{17} \times \frac{21}{50}$	34	$\frac{3}{10}$	35	$\frac{6}{35}$
36	$\frac{16}{45}$	37	$\frac{6}{11}$	38	$\frac{2}{7}$	39	$\frac{7}{48}$
40	$\frac{5}{12}$	41	$\frac{5}{12}$	42	$\frac{8}{21}$	43	$\frac{1}{2}$
44	$1\frac{1}{5}$	45	$2\frac{1}{4}$	46	$6\frac{2}{3}$	47	$16\frac{1}{2}$
48	$10\frac{1}{2}$	49	12	50	$\frac{1}{2}$	51	$1\frac{1}{4}$
52	$\frac{1}{125}$	53	3p	54	$\frac{14}{15}$	55	$1\frac{1}{5}$
56	3	57	2	58	$2\frac{4}{9}$	59	$2\frac{2}{3}$
60	1	61	$\frac{4}{9}$	62	0.5	63	0.4
64	0.375	65	0.35	66	0.6	67	0.17

13 Infusion rates (page 28)

1 **a** 2 **b** 8 **c** 1 **d** 100 **e** 400 **f** 2000 **g** 10 000
2 **a** 14 **b** 16 **c** 7   3 **a** 27 **b** 17 **c** 42
4 30   5 50   6 16   7 2
8 5   9 7 (6.4)   10 38   11 25
12 **a** 2 hrs 30 mins **b** 3 hrs 20 mins **c** 6 hrs 40 mins

14 Ratio (page 30)

1 9   2 $\frac{4}{9}, \frac{5}{9}$ 3 Residential £20 000, Day £4000
4 Green £200 000, Blue £300 000, Red £500 000
5 **a** 2 in 5 **b** 1 in 6 **c** 7 in 12 **d** 3 in 13
6 **a** 2:1 **b** 3:5 **c** 2:5 **d** 4:5
7 50 ml stock, 450 ml water
8 **a** $\frac{1}{4}$ **b** 250 ml
9 **a** 300 ml **b** (i) 1200 ml (ii) 1.2 litres
10 2 litres   11 3:4   12 1:2
13 Elwood £30 000, Breedon £25 000   14 3:5:7

15 Percentages (page 33)

1 15%   2 25%   3 33.33%   4 51%
5 25%   6 10%   7 5%   8 48%
9 1.2%   10 71.42%   11 71.42%   12 66.66%
13 5%   14 8.33%   15 75%   16 50%
17 45.45% 18 2.5%   19 66.66%   20 25%

16 Percentages, fractions and ratios (page 35)

1 $\frac{1}{10}$ 2 $\frac{7}{20}$ 3 $\frac{17}{100}$ 4 $\frac{9}{50}$
5 $\frac{3}{40}$ 6 $\frac{9}{400}$ 7 **a** 1 in 25 **b** 1:24
8 **a** 2 in 25 **b** 2:23   9 **a** 1 in 8 **b** 1:7   10 **a** 3 in 40 **b** 3:37

17 Solution strength (page 36)

1 200 ml   2 30 ml   3 125 ml
4 **a** 40 ml **b** 1960 ml   5 **a** 200 ml **b** 4.8 litres
6 **a** 200 ml **b** 800 ml   7 40 ml
8 10 ml   9 **a** 75 ml **b** (i) 1425 ml (ii) 1.425 litres

18 Percentage problems (page 38)

1 **a** 12% **b** 0.25%   2 68%   3 15.15%
4 **a** 0.2% **b** 0.13%   5 6.43%   6 12.62%
7 **a** 67.27% **b** No, discuss   8 25%

9   **a** 0.14%   **b** 0.29%   10   83.33%   11   10.24%
12 **a** 25%   **b** 0.25%
13 **a** No   **b** 54%   14   95%   15   4.08%

19 Puzzles (page 40)

2   $\frac{119}{7}$ 3 $\frac{5}{200}$ 4 $\frac{24}{6}$
5   **a** £272   **b** £1700   6   **a** 2.5 ml   **b** 4.375 ml
7   400   8   **a** 0.33 ml **b** 0.42 ml **c** 0.6 ml
9   **a** 400 ml   **b** 0.4 litres
10 £3   11   9   12 £880 000

20 More about percentages (page 42)

1   £3.00   2   8 ml   3   £4.75   4   50 g
5   £0.26   6   4.05 m   7   £17   8   18 g
9   £87.50   10 750 ml   11 £0.04 (4p)   12 1.6 kg
13 £690   14 30 ml   15 140 mg   16 180 m
17 5 litres   18 £12.50   19 135 g   20 210 ml

21 More percentage problems (page 43)

1   **a** 20 ml   **b** 80 ml   2   £57   3   £990 000
4   £495   5   **a** £1.65   **b** £7.15   6   15
7   360   8   £68.80
9   Present investment by £700 per year – depends on types of investments and outlook
10 £1375

22 Profit and loss (page 44)

1   **a** £1580   **b** 75.24%   2   **a** 31.47%   **b** 12%
3   **a** One hundred million, two hundred thousand pounds   **b** £6 012 000
4   £8000 loss   5   £1770

23 Costing a meal (page 46)

1   £0.54   2   £0.62   3   £1.32   4   £0.16
5   **a** £0.25   **b** 25%   6   10%   7   60%   8   75p
9   £1.68   10 60p   11 £1.70   12 40p

24 Preparing a budget (page 49)

3   **a** £920   **b** £886   **c** Fees, meals, wages, food and drink
4   **a** (i) £50 000 (ii) 10%   **b** (i) £46 500 (ii) 9.3%

25 Stock control (page 51)

1   6 towels   2   12 tins   3   £9   4   £19.60
5   Deterioration (stock life), capital tied up, storage space, obsolescence, demand, falling prices
6   Reliable supplier, demand, rising prices

26 Graphs (page 53)

1   37.4°C   2   Evenings of 5 April, 8 April
3   36.8°C   4   37.4°C   5   £200   6   Weeks 2, 7, 10
7   Weeks 8, 9   8   £200   9   Graph   10 Graph
11 4.4 lbs   12 1.8 kgs   13 8.8 lbs   14 Graph
15 £8 000 000   16 Substance misuse
17 Domestic violence   18 Chart   19 Chart

27 Pie charts (page 57)
1 £100 000   2 160°   3 80°, £400 000   4 Chart
5 60°   6 90°

28 Centigrade and fahrenheit (page 58)
1 a 50°F   b 113°F   c 98.6°F
2 a 68°F   b 32°F   c 100.4°F
3 a 15°C   b 100°C   c 35°C
4 a 194°F   b 95°F   c 113°F
5 a 5°C   b 80°C   c 25°C

29 Computer spreadsheets (page 60)
1 $B \div A \times 100$
2 a $A - B$ b $B \div A \times 100$
3 $B \div A \times 100$
4 $(A - B) \div A \times 100$

Test 1 (page 63)
1 a 131.115   b 1.805 kg
2 a $\frac{2}{3}$ b $\frac{2}{3}$
3 a £63.75   b Gross is before any deductions, Net is after deductions for tax, insurance, pension, etc. (take home pay)
4 15 sq m   5 a 13 050 ml   b 5 lbs 3 oz   6 12
7 7 days   8 £18.81   9 0.05 ml   10 10:17

Test 2 (page 64)
1 a $17\frac{1}{10}$ b 0.25
2 a 0.12   b 12%   3 30.8 lbs
4 £2775 : £1850 : £925   5 3575 ml   6 250 kg
7 Balance £2.88   8 When total receipts = total expenditure
9 £19.42   10 a negative   b (i) 0.2 litres (ii) 200 ml

Test 3 (page 65)
1 a £123.25   b 10.625 kg   2 9.74   3 £8513.40
4 6   5 340 ml   6 58.82%   7 £6.25
8 Increase   9 a below   b minus   10 600

Test 4 (page 66)
1 a 22.95 pints   b 3 gallons   2 a $1\frac{7}{8}$ b 109
3 Halton £250 000, Hove £200 000, Hanley £150 000, Huncote £400 000
4 400 ml   5 £115.50   6 67.56%   7 3
8 £35 000   9 36   10 a £9400, 26.11%   b profit

Test 5 (page 67)
1 a 3.774 litres   b 3774 ml   2 3563 g
3 a 59°F   b 10°C   4 Graph   5 90.24%
6 110   7 £96.72   8 585 mg
9 a 10.6 sq ft   b Yes   10 a 3.505 kg   b 8 lb

Test 6 (page 68)
1 a $\frac{18}{107}$ b $\frac{13}{17}$ 2 0.85   3 Balance £9.86
4 a 1.25 ml   b 19   5 2045 francs   6 £73.15
7 Greensleaves £2050, Knighton £4100, Rowlett £8200
8 a 4.17 amps   b 5 amps
9 Low stock – less money tied up, less storage space required, less chance of deterioration, security easier and insurance rates lower
10 a 16.55 hrs   b 4.55 p.m.

Test 7 (page 70)

1   **a** 14 540 mm   **b** 0.05 kg   2   40%
3   **a** 30   **b** Darker evenings – G.M.T.   **c** 60   **d** Yes
4   **a** 69.44%   **b** £7875   **c** £11 340   **d** £147 420
5   252   6   **a** 56 units   **b** 0.56 ml   7   size 14
8   **a** 4547   **b** 5%   9   **a** £1.11   **b** £2.47   **c** 55.6%
10   £3386 excess of income over expenditure

Test 8 (page 71)

1   **a** two million, three hundred thousand and fifty-seven   **b** 902 200
2   **a** 93.6   **b** 836 kg
3   34.94%
4   **a** 120 ml   **b** 1880 ml
5   **a** (i) 1.675 m (ii) 58.18 kg   **b** Yes
6   **a** (i) 12.16% (ii) 63.06%   **b** 1 in 37   **c** Not every alarm was used
7   London £53 000, Cornwall £6 649 000, Lancashire £20 081 000, Kent £19 198 000, Isle of Wight £2 314 000, Suffolk £6 135 000
8   £7.20   9   £2.68
10   **a** $(A - B) \div A \times 100$
     b Numbers of girls who have had German Measles and their % relating to A

Test 9 (page 73)

1   **a** 34.2   **b** 82.26   2   71   3   5   4   30 ml
5   Capital tied up especially when prices are falling, deterioration, storage capacity, out-dated
6   11 October   7   **a** chart   **b** 19°
8   **a** £1058.75   **b** £974.05   9   **a** 2400 cals   **b** 330g
10   70

Test 10 (page 75)

1   **a** 1.325 litres   **b** 1325 ml   2   **a** 1 750 000   **b** 250
3   **a** 86 500   **b** 850 000   4   **a** $\frac{5}{13}$ **b** 38.46%
5   **a** 18.43   **b** 4.30 p.m.   6   £19.87
7   **a** 7.77 mins   **b** 15.38%   8   £1656
9   **a** Sandra £5400, Graham £3326.40   **b** Sandra by £2.27 per hour (approx.)
10   **a** 60%   **b** 3:2

Test 11 (page 76)

1   **a** 14.23   **b** £2.24   2   928 g
3   **a** £79.95   **b** (i) 0–2 yrs (ii) 1
4   **a** £160   **b** (i) £12.50 (ii) £20.00   **c** 20.8%   **d** Mr Brown (discussion)
5   **a** £43.52   **b** £63.45   6   **a** 13%   **b** 22p
7   **a** 6 (to nearest whole number)   **b** 4   **c** 4   8   10.44%
9   Swithland – £720 000, Charnwood – £600 000, Cropston – £480 000
10   £1000

Test 12 (page 79)

1   **a** $13\frac{4}{7}$ **b** $\frac{5}{6}$ 2   £3 330 000 000
3   **a** graph   **b** 16 inches   **c** 535 mm   4   £14 000
5   1.25 ml   6   **a** £335 252.50   **b** £5.96   7   Balance £2.38
8   **a** £800   **b** £1650   **c** £780   **d** 5%   **e** £2780
9   **a** 414%   **b** 60.6%   **c** 9000
10   **a** 3.014 kg   **b** 3 kg   **c** 2.9 kg

Index

Accounts 20
  personal 20
  petty cash 22
Addition 1
Answers 82
Area 12
Averages 15
  mean 15
  median 15
  mode 15

Balancing accounts 21
Bar charts 56
Bin card 51
Block graph 56
Blood temperature 53, 59
Boiling point 59
Break-even point 44
Budgeting 49
  proforma 50

Cancelling 24
Capacity tables 9
Capital 45
Cash accounts 20
Centigrade 58
Computer spreadsheets 60
  disks 61
  hardware 60
  programmes 60
  software 60
Correcting 7
Costing a meal 46
  proforma 46

Days inclusive 17
Delivery note 51
Disks – computer 61
Division 5
  fractions 27
  tables 5
  quick methods 8
Drip rates 28

Fahrenheit 58
Float of petty cash 22
Fluid balance chart 65
Food costs 46
Formulae
  electrical 69
  solution strength 36
  spreadsheets 61
  temperatures 59
Fractions 24

  cancelling 24
  division 27
  mixed numbers 25
  multiplication 25
Freezing point 59

Graphs 53
  bar 56
  block 56
  histogram 56
  line 53
  straight line 55
  temperature conversion 56
  weight conversion 55

Hardware – computer 60

Imperial measurement 8
Imprest petty cash book 22
Improper fractions 25
Inclusive days 17
Infusion rates 28
Ingredients 46
Intravenous set 28

Length tables 9

Mean 15
Median 15
Metric measurement 8
Mixed numbers 25
Mode 15
Multiplication 4
  fractions 25
  quick methods 5
  tables 4

Percentages 33, 35, 42
  problems 18, 43
Petty cash 22
Pie charts 57
Pricing
  meal 48
  fees 80
Profit and loss 44
Programming – computer 60
Protractor 27
Puzzles 40

Ratio 30, 35

Selling price (food) 48
Sharing 30
Software – computer 60
Solution strength 36
Spreadsheets 60

Stock control   51
  bin card   51
  deterioration   51
  rotation   51
  valuation   77
Subtraction   3

Tables
  division   5
  further   10

  measurement   9, 10
  multiplication   4
Temperature   58
  blood   53, 59
  boiling point   59
  freezing point   59
Tests   63
Twenty-four hour clock   19

Weight tables   9, 10

just THE JOB Law & Order

Lifetime Careers
WILTSHIRE

Hodder & Stoughton

A MEMBER OF THE HODDER HEADLINE GROUP

Just the Job! draws directly on the CLIPS careers information database developed and maintained by Lifetime Careers Wiltshire and used by almost every careers service in the UK. The database is revised annually using a rigorous update schedule and incorporates material collated through desk/telephone research and information provided by all the professional bodies, institutions and training bodies with responsibility for course accreditation and promotion of each career area.

ISBN 0 340 68780 0
First published 1997

Impression number   10 9 8 7 6 5 4 3 2 1
Year   2002 2001 2000 1999 1998 1997

Copyright © 1997 Lifetime Careers Wiltshire Ltd

All rights reserved. No part of this publication may be reproduced or transmitted in any form or by any means, electronic or mechanical, including photocopy, recording or any information storage and retrieval system, without permission in writing from the publisher or under licence from the Copyright Licensing Agency Ltd. Further details of such licences (for reprographic reproduction) may be obtained from the Copyright Licensing Agency Ltd, 90 Tottenham Court Road, London W1P 9HE.

Printed in Great Britain for Hodder & Stoughton Educational, the educational publishing division of Hodder Headline Plc, 338 Euston Road, London NW1 3BH, by Cox & Wyman Ltd, Reading, Berkshire.

CONTENTS

Introduction	7
Opportunities in law outside private practice	11
Lawyers in industry and commerce. Lawyers in the Civil Service. The Crown Prosecution Service. Lawyers in local government. Teaching. International opportunities. Other opportunities. Non-legal work.	
Solicitor	15
Legal executive	19
Barrister	24
Barristers' clerk	31
Legal secretary	34
Licensed conveyancer	38
Working in court services	41
Justices' clerk (clerk of court). Justices' chief executive. Justices' clerk's assistant. Financial staff. Usher. Court reporting. Coroner's officer. Bailiff.	
Judges & magistrates	47
Registrar of births, marriages & deaths	50
Registrar. Superintendent registrar.	
Coroners & coroners' officers	53
Coroner. Coroner's officer.	
The prison service	56
Prison officer. Governor. Specialist jobs.	
Security work	60
Police work	64
Traffic police. The CID (Criminal Investigation Department). Crime prevention officers. Dog-handlers. Mounted police. River police. Underwater units.	

Armed police. British Transport Police. Special Constables. Ministry of Defence police. Administration, clerical and computing jobs. Traffic warden. Law clerk. Scene of crime officer.

Private detective work	**73**
Forensic science	**75**
Immigration, customs & excise	**78**
HM Immigration Service. Customs and Excise	
Patents & trade marks	**80**
Patent agent. Patent examiner.	
Company secretary	**85**
For further information	**90**

JUST THE JOB!

The ***Just the Job!*** series ranges over the entire spectrum of occupations and is intended to generate job ideas and stretch horizons of interest and possibility, allowing you to explore families of jobs for which you might have appropriate ability and aptitude. Each ***Just the Job!*** book looks in detail at a popular area or type of work, covering:

- ways into work;
- essential qualifications;
- educational and training options;
- working conditions;
- progression routes;
- potential career portfolios.

The information given in ***Just the Job!*** books is detailed and carefully researched. Obvious bias is excluded to give an even-handed picture of the opportunities available, and course details and entry requirements are positively checked in an annual update cycle by a team of careers information specialists. The text is written in approachable, plain English, with a minimum of technical terms.

In Britain today, there is no longer the expectation of a career for life, but support has increased for life-long learning and the acquisition of skills which will help young and old to make sideways career moves – perhaps several times during a working life – as well as moving into work carrying higher levels of responsibility and reward. ***Just the Job!*** invites you to select an appropriate direction for your *own* career progression.

Educational and vocational qualifications

A level – Advanced level of the General Certificate of Education
AS level – Advanced Supplementary level of the General Certificate of Education (equivalent to half an A level)
BTEC – Business and Technology Education Council: awards qualifications such as BTEC First, BTEC National Certificate/Diploma, etc
GCSE – General Certificate of Secondary Education
GNVQ/GSVQs – General National Vocational Qualification/ General Scottish Vocational Qualification: awarded at Foundation, Intermediate and Advanced levels by BTEC, City & Guilds, Royal Society of Arts and SQA
HND/C – BTEC Higher National Diploma/Certificate
International Baccalaureate – recognised by all UK universities as equivalent to a minimum of two A levels
NVQ/SVQs – National/Scottish Vocational Qualifications
PGCE – Postgraduate Certificate in Education
SCE – Scottish Certificate of Education, at **Standard** Grade (equate directly with GCSEs: grades 1–3 in SCEs at Standard Grade are equivalent to GCSE grades A–C) and **Higher** Grade (equate with the academic level attained after one year of a two-year A level course: three to five Higher Grades are broadly equivalent to two to four A levels at grades A–E)

Vocational work-based credits	NVQ/SVQ level 1	NVQ/SVQ level 2	NVQ/SVQ level 3	NVQ/SVQ level 4
Vocational qualifications: *a mix of theory and practice*	Foundation GNVQ/GSVQ; BTEC First	Intermediate GNVQ/GSVQ	Advanced GNVQ/GSVQ; BTEC National Diploma/Certificate	BTEC Higher National Diploma/Certificate
Educational qualifications	GCSE/SCE Standard Grade pass grades	GCSE grades A–C; SCE Standard Grade levels 1–3	Two A levels; four Scottish Highers; Baccalaureate	University degree

INTRODUCTION

There is a broad range of job opportunities available within the various branches of the legal profession. Some of the jobs involve working for the government while others are in private firms. Not all law jobs are concerned with crime and law-breaking. There is lots of work in family law, house-buying, wills and divorce, for example, and in business and industry dealing with contracts, patents and trade marks. Qualifications vary, but solicitors and barristers are virtually always graduates. Some jobs are open to people with good school exam passes.

The legal system in England and Wales differs from those in Scotland and Northern Ireland. There are differences in law, in how the courts are organised, and also in the training and organisation of those working in the legal profession. Career options are therefore normally limited to the legal system in which you have trained (see the Further Information section for relevant addresses).

Would I like it?

- Can you get on with all different kinds of people?
- People representing the legal system need to look smart. Barristers wear wigs and gowns in court, and court ushers wear robes.
- You will need to be able to work on your own, but also as part of a team.
- You may have to work under pressure to meet deadlines and/or targets.
- For most jobs you will have to keep studying throughout your career, as law changes all the time.

- You will have to be able to understand and explain complicated legal matters.
- Some court cases can be very upsetting. Could you cope without getting too involved?

Solicitor

Solicitors are the people in the legal profession who have regular contact with the public, offering advice and assistance on all kinds of legal matters. They may also represent their clients in magistrates' courts and county courts, but not normally in the higher courts. They need a wide knowledge of the law, but, in a large practice, they may be specialists in areas like company law or divorce and custody cases. The usual pathway to becoming a solicitor is via a law degree, followed by further training, although there are alternative routes.

Legal executive

Legal executives work for solicitors, carrying out the more routine, but often quite specialised, legal work. With experience, legal executives are able to carry out work which demands a high level of responsibility. They may handle the legal side of property deals or they may draft wills. Four GCSEs at grade C or above, including English language, and sometimes A levels or qualifications of an equivalent standard, are required to get a position as a trainee legal executive in private practice or public service.

Barrister

Barristers work in the higher courts, defending and prosecuting cases which have been referred to them by solicitors. They may also represent clients at other proceedings of a legal nature, such as public enquiries. Some barristers work as legal advisers and consultants. From the ranks of barristers are selected Queen's Counsels and judges. To become a barrister, it is necessary to have at least a second-class honours degree, preferably in law.

Barristers' clerk

Barristers' clerks organise the running of barristers' chambers where a group of barristers practise. Opportunities for work are limited to London and a few large centres. Entry qualifications are a minimum of four GCSEs at grade C, to include English and maths, or their equivalent.

Magistrates' courts services

Justices' clerks are the qualified solicitors or barristers who manage the magistrates' courts services and provide legal advice to magistrates. Magistrates are part-time officials, with the exception of those working in London. The clerk's assistants who carry out the routine work of the court, and advise magistrates, are trained on-the-job. You need several good GCSE passes – in practice A levels or their equivalent may be required – to get a trainee post. There are also supportive jobs in coroners' courts.

Crown Prosecution Service

The Crown Prosecution Service prosecutes offenders on behalf of the police, and is part of the Civil Service. Prosecutors are civil servants and also qualified solicitors or barristers. The service is the largest employer of lawyers in the country.

Court reporting

Court shorthand writers or stenotypists, often referred to as verbatim reporters, may be employed to take down all the court proceedings, normally using a computerised system, to provide an accurate account of what occurs. They are not employed by courts but by firms specialising in this work. Sometimes they are freelance writers.

Licensed conveyancer

Conveyancers deal with the legislation and administration involved in transferring property or land from one owner to another. They may work with or for solicitors, estate agents or other conveyancers. To start training as a licensed conveyancer, you need a minimum of four GCSEs at grade C, including English language, or equivalent qualifications.

Legal secretarial and administrative work

Legal secretaries and legal cashiers and administrators are especially trained to work in solicitors' offices and other legal environments. They deal with the preparation of such documents as wills, divorce petitions and witness statements, as well as undertaking the usual secretarial duties.

The European scene

Law graduates are being recruited into the 'European Fast Stream' scheme to encourage more British lawyers into European Commission work. Students who are accepted for the scheme are sponsored during their vocational year, and offered articles in different departments of the Government's Legal Services. Entry depends upon achieving a first or upper second class degree and being selected through undertaking Civil Service tests. Posts in Europe are not guaranteed; applicants must pass the European Commission entrance procedures.

The following sections give fuller information about these and other career opportunities in the law profession itself, as well as jobs in law-related areas such as the prison service, police and security work, and customs and excise.

just THE JOB

OPPORTUNITIES IN LAW OUTSIDE PRIVATE PRACTICE

> Many newly trained and qualified solicitors and barristers go into private practice – the area of the profession where 90% of trainee solicitors complete their training contracts. However, there are other career openings for members of both branches of the legal profession. Many positions for lawyers are to be found in industry and commerce, the Civil Service and local government.

Lawyers in industry and commerce

Large organisations, such as the former nationalised industries, building societies and banking companies, and big concerns like Shell, ICI and the larger engineering and building firms, all have legal departments. The work covers contracts and matters concerning company law, property conveyancing and property development work, tribunals and race and sex discrimination law. There have been recent changes of legislation in all these areas – which are proving to be a growing source of work for lawyers.

Lawyers in the Civil Service

The government employs lawyers who provide services to a wide range of departments and government agencies. Some are engaged in litigation and conveyancing in much the same way as private practice solicitors. Others are concerned with the government's legal powers, constitutional matters, drafting legislation, the work of the office of the Director of Public Prosecutions and even such things as the Charity Commissioners. Approximately 1000 solicitors are employed in government legal services.

The Crown Prosecution Service

This is a department of the Civil Service, headed by the Director of Public Prosecutions. It is responsible for the conduct of all criminal proceedings instituted by the police, except for the most minor offences, thus playing a major role in the criminal justice system in England and Wales. The Crown Prosecution Service is the largest employer of lawyers in the country, with over 2000 engaged in active work.

There are also many work opportunities for administrators without legal qualifications within the Crown Prosecution Service.

Lawyers in local government

Lawyers in local government deal with all aspects of work in the community, ranging from housing to social services, planning to transportation, environmental health to highways – depending on the type of local authority. Lawyers advise individual departments within the Council, but also advise elected members and committees on local government legislation and on their duties and powers within the law. Because local authorities are landholders and employers, there are also numerous commercial, property and employment law matters to be attended to. About 3000 solicitors are employed within local government departments at the present time.

Advocacy plays a large part in this work – including presentations where the Council is the regulatory body, civic litigation on contracts, property and debt, and planning inquiries and industrial tribunals.

Local authorities offer training contracts to trainee solicitors, and provide an extensive training programme which includes advocacy experience. Many chief executives of local authorities started their careers as lawyers.

Teaching

Some law graduates become teachers, primarily in colleges of further or higher education and universities. Besides training the next generation of law students, they also teach students on courses where an appreciation of the law is necessary – for example, in the field of business studies or company secretaryship. Some law graduates are engaged in teaching people entering the police force.

International opportunities

Lawyers of many nationalities work at the World Court and the European Courts, or for the European Commission. The Civil Service operates a 'European Fast Stream' scheme for trainee solicitors and pupil barristers.

Multinational companies may offer prospects for lawyers who want to travel. Some further training or accreditation may be required, and the ability to speak other languages is obviously an asset.

The Law Society produces a range of leaflets and booklets on the international legal field – see address in Further Information section.

Other opportunities

Other opportunities which could need a specialist knowledge of law include work with accountancy firms, academic or other special librarianship work, journalism, work in community law advice centres, and the police service.

Non-legal work

A legal qualification, including a degree, can be a useful starting point from which to enter the general field of graduate employment.

just THE JOB

SOLICITOR

> Solicitors are specialists on all matters of law to the public. They are generally the first point of contact for those with a legal problem, providing them with skilled legal advice, and representing them in court when necessary. Most solicitors work in private practice. To become a solicitor, you need three academic A levels with good grades, followed by a degree course and postgraduate study.

What it takes

- Could you stand up in a magistrates' court to defend a client accused of drunken driving?
- Have you the patience to read and master a mass of complicated financial papers in a fraud case?
- Could you calmly advise one party in a very bitter divorce case?
- Could you sort out the affairs of a wealthy client who had died without making a will and hadn't spoken to his relatives for twenty years?

These are the sort of problems you may well have to deal with as a qualified solicitor: they are often concerned with people in situations requiring tact and sensitivity. Part of the solicitor's job is making the law comprehensible to clients and guiding them through the legal maze of courts, oaths, witnesses, counsels and robed, bewigged judges.

The legal system in England and Wales is conducted by members of two professions – solicitors and barristers. As a solicitor you can represent clients in a magistrates' court. In a higher court, a barrister is generally engaged to give specialised advice and to present the case in court.

There are over 60,000 solicitors in England and Wales, working in private practices, business, industry, and local and central government. Individual private practices may specialise in areas such as criminal defence or company law, or they may offer a comprehensive service. Within a practice, solicitors may be specialists in areas such as matrimonial matters, property, media and entertainment, personal injury, intellectual property or immigration law.

EDUCATION AND TRAINING

The training of solicitors is a long process. Progress from student to fully-fledged solicitor will take at least six years. Students need to enrol with the Law Society approximately six months before starting their postgraduate training course. There are various entry routes which you can take into the profession, depending on your qualifications and experience.

Taking a qualifying law degree

This is the usual way into the profession, especially if you are convinced that the profession is for you. A list of qualifying degree courses is available from the Law Society. Some degrees combine law with subjects like languages, computing or economics. As long as the course covers the six core subjects of constitutional and administrative law, law of contract, law of tort, criminal law, land law, and equity and trusts, it may still be listed as a qualifying degree.

The minimum entry qualifications are two A levels, with supporting GCSEs including English language, but generally, three good A levels are required. A level law is certainly not required.

You would then follow these three stages:

- a three- or four-year degree course;
- a one-year full-time, or two-year part-time, Legal Practice course;
- two years' training contract in a solicitor's office or other approved organisation. There are not enough training contracts available to meet the demand from applicants, so this stage of qualifying is very competitive.

Taking a non-law degree

With a degree in any subject other than law, you would then follow these three stages:

- either a one-year full-time or two-year part-time course leading to the Common Professional Examination, or a postgraduate Diploma in Law;
- a one-year full-time, or two-year part-time, Legal Practice course;
- two years' training contract, as above.

There are one or two possibilities of gaining the Diploma by distance learning. It is unusual to get a grant for any of these stages.

Non-graduate entry

It is possible for a school-leaver or mature non-graduate to train as a solicitor after first qualifying as a Fellow of the Institute of Legal Executives and serving five years in a solicitor's office or similar workplace. This is a longer but cheaper training route, taken by about 200 people each year.

Continuing education

All solicitors are now expected to undertake continuing education to update themselves in revisions and additions to the law.

GRANTS

Grants for degree courses are mandatory, but all other law courses are only eligible for discretionary award funding. You may find yourself unable to obtain finance to complete the necessary courses. Contact your careers office or the Awards Section of the local education authority to check the current position.

The Law Society publishes a leaflet about funding for Common Professional Examination, Diploma and Legal Practice courses, giving information on bursaries and sponsorships which may offer a limited source of financial help.

LEGAL EXECUTIVE

> Legal executives are professionally qualified to follow a career in the law, working alongside solicitors, helping their clients to get results. Due to the very broad nature of a solicitor's workload, it is often impossible for him or her to handle all their work in person, so most solicitors employ one or more legal executives to help them. You need at least four GCSEs at grade C, or the equivalent, to commence training – although some employers now expect qualifications of A level standard.

The work of legal executives is very similar to that of solicitors and quite wide-ranging. They have a wide knowledge of law, but generally are specialists in one particular aspect. They may work in:

- **conveyancing** – the sale, purchase and leasing of land and buildings;
- **litigation** – preparation for court cases: legal executives interview clients and witnesses, prepare statements of evidence, etc;
- **probate** – dealing with the property of people who have died: attending to the valuation of property and carrying out the terms of a will;
- **accountancy** – dealing with clients' money, keeping accounts and managing investments.

Legal executives undertake a lot of practical, detailed work on cases – preparing documents, doing research into a particular case and so on. Some court work could be involved – appearing before Masters of the High Court or District Judges in the County Court, or in chambers, and some legal executives are entrusted to take on caseloads of their own.

Most legal executives work with solicitors in private practice, but about 20% work for the legal departments of large organisations such as government, local authorities, commerce and industry.

What it takes

- good communication skills – legal executives have to deal with people, interviewing clients and witnesses, and must be able to put people at their ease: they have to be able to explain points of law clearly;
- honesty and discretion;
- concise and logical thought;
- a liking for detail;
- an ability to cope under pressure.

Kath – legal executive

Most people know roughly what a solicitor and a barrister do, but they often look a bit puzzled when I tell them what my job is. I am really a sort of assistant solicitor and the name legal executive is an official legally protected name. You can't call yourself a legal executive unless you have passed exams set by the Institute of Legal Executives.

I work in a solicitor's office on personal injury disputes. Where somebody has been injured in a car accident and thinks that their injury is the result of, perhaps, careless driving, I can help them to use the legal system to get compensation. Of course, I can also be working on the other side, defending somebody like an employer who is being sued by an employee.

I have to have a very wide knowledge of this part of the law, and as a specialist, I cover all this type of work for our firm. I see clients, interview possible witnesses, draw up letters and, in some cases, even make court appearances, though this isn't a major part of my job. Lots of the cases I work on never even get as far as a court. The dispute is settled by an agreement between the people involved, which is often much cheaper and easier for all concerned.

I left school with five GCSEs and one A level, without any idea of working in the law. After a short period working in a building society, I saw an advert for a trainee legal executive post with a solicitor. I've just finished four years of training, with evening classes and day release. My firm has been very helpful, giving me time off to study and help with buying all the books I needed. Now I'm qualified, I'm going to take a break from exams. In the future, I might think about trying to qualify as a solicitor, but for the moment I'm very happy with what I'm doing.

QUALIFICATIONS

There are no formal specified qualifications which are needed to obtain work in a solicitor's office. Some employers will recruit trainee legal executives who have four academic GCSEs at grade C, including English, or the equivalent, but it is now more common for trainees to hold A levels, BTEC National or Advanced GNVQ qualifications. However, many of the people who eventually qualify as legal executives start as legal secretaries or clerks, with fewer educational achievements.

The minimum qualifications needed to register with ILEX (the Institute of Legal Executives) to follow the legal executive training scheme are four GCSEs at grade C (or the equivalent), including English language and three academic subjects such as economics, geography, history, languages, mathematics, sciences, religious knowledge, sociology, commerce or accounts. Applicants who want to take the legal executive training scheme but do not have the necessary qualifications, can take either the ILEX Preliminary Certificate in Legal Studies, a one-year course, or one of a number of ILEX vocational legal qualifications which are designed for people who are not working in solicitors' offices.

Entrants with A levels are welcomed as trainees, as it is a suitable career for someone interested in law but who cannot obtain sufficiently high A level grades to train as a solicitor in the usual way. At the present time, about 15% of entrants are graduates.

Applicants over the age of 21 may enrol as mature students with the backing of a reference letter from their employer or a suitable professional person.

TRAINING

To qualify to use the title 'legal executive', you need to become a Fellow of the Institute of Legal Executives. To become a Member involves taking Parts I and II of the Institute's Membership Examination. Most people are employed in a solicitor's office while studying part-time for the exams, but you can take both parts without working for a solicitor. There are a few full-time courses available. Both parts of the exam can be studied by distance learning.

You can apply to become a Fellow of the Institute when you are 25 years of age and have completed five years in a solicitor's office, at least two of them after passing the Membership examination.

Going on to become a solicitor

The Law Society states that a Fellow (by examination) of the Institute of Legal Executives has the required standard of general education to train as a solicitor, and has earned exemption from some examinations.

Fellowship gives exemption from three of the six core subjects of the Common Professional Examination, and the other three may be taken by part-time study. The next step is to take the Legal Practice Course, studying full-time for one year or part-time for two years, followed by the Professional Skills Course. These two courses are the final stages of examination and training for qualification as a solicitor. Fellows are exempt from the Training Contract period.

Qualified Members of the Institute may also apply for exemption from the Common Professional Examination and to join a Legal Practice Course, provided they have had at least three years' qualifying experience in legal work after attaining the age of 18. However, members of ILEX taking this route will be required to enter into a training contract.

Adults: note that maturity and previous experience may mean that stated entry requirements can be relaxed for people over 21 years of age.

just THE JOB

BARRISTER

> In England and Wales, the legal profession is divided into two branches of practising lawyers – barristers and solicitors. Solicitors advise members of the public on legal matters. Barristers do not have direct contact with the public, but provide advocacy services (presenting cases in court) and give expert specialist legal opinion to solicitors and their clients. All barristers are graduates who then undergo further training.

The relationship between solicitor and barrister is similar to that between family doctor and consultant in medicine. There are more than five times as many solicitors as barristers. About two-thirds of all barristers have chambers in London; the rest are based in other cities throughout England and Wales.

Barristers are:

- primarily concerned with serving the community, both in high court and in the community courts;
- consulted on points of law, and may conduct some research before giving an opinion;
- consulted by solicitors for an opinion of whether a case is likely to succeed should it come to court;
- involved with drafting written opinions of legal documents for solicitors and their clients.

Although recent legislation has extended the provision of legal services, and solicitors can now acquire the qualifications necessary to appear as advocates in the higher courts, it is still barristers who present the majority of cases of any importance to a court of law. They draft the pleadings and other documents necessary in a particular case and advise on the evidence needed by their client. Barristers are, in particular, specialists in advocacy – which is to present a case in court before a judge. Advocacy involves learning to sift evidence to assess the weight of conflicting statements and learning to carry out cross-examinations. A client's case must be presented in the best possible way, but the barrister must also assist the court in determining the truth of the matter, so that justice may be done.

Aside from court work, some barristers are employed in commerce and industry, acting as legal advisers on subjects such as company law, patents, mergers, etc. They are also employed in central and local government, providing an advisory service to civil servants and ministers. Employed barristers do not usually appear as advocates in court, although in certain circumstances they may do so.

What it takes

As the lengthy training indicates, you need to acquire a good many skills before becoming a successful, practising barrister. These include:

- a high level of intellectual ability to take in facts, analyse them and act upon them;
- a good command of language, to present a case clearly and fluently, and cross-examine effectively;
- excellent presentation skills, and powers of persuasion;
- self-confidence;
- honesty;
- mental and physical stamina – cases may be complex and stressful, and you may remain in the public gaze for a considerable length of time.

Sarah – barrister

❝ I have been working as a fully qualified barrister for three years now, and I am really only just getting established. The training is very long and expensive, and it's hard to get through without having to borrow money. The idea is that, once you're working, you will eventually earn enough to pay back the loans. The prospects are very good and the court-room atmosphere can be really exciting.

I started off by taking an English degree, and then had to do another two years' legal training, followed by what is called pupillage, where you work as a trainee for a year in a barrister's chambers.

I am fortunate enough to be able to earn some regular money as a legal reader for a newspaper, checking that they are not publishing libels, which helps to pay the bills!

Of course, a senior barrister can work on important cases and earn very high fees, but a junior like me has to take whatever comes along. I work with a group of barristers in a set of chambers in London, where we take both criminal work and civil cases. I'm experienced enough now to conduct straightforward cases myself, but on a complex case I assist more senior colleagues.

Appearing in court was very alarming at first. I was terrified the judge would tell me to go away and put my wig on straight (they have been known to do this). But of course, the important thing is to be well prepared, which can mean burning the midnight oil before a court appearance. I have to put in a lot of work just to keep up with all the changes caused by new laws and judgments made in court. Winning a case is like winning a debate or an argument: it can be very rewarding. ❞

EDUCATION AND TRAINING

The four Inns of Court have always had an interest in the education and training of barristers. For centuries this was conducted in the Inn halls, where members took their meals in common, and students are still required to dine in the Inn a certain number of times during their training. The Council of Legal Education supervises the vocational training of barristers at the Inns of Court School of Law in Gray's Inn and runs the Bar Vocational Course and examinations. Students can apply for membership of an Inn of Court early in the final year of their degree studies at university.

Some chambers offer 'mini-pupillages' for undergraduate would-be barristers to experience a taste of chamber life before embarking on their postgraduate training.

The Bar Vocational Course for would-be barristers is now available at a number of institutions other than the Inns of the Court School of Law. However, although there will be more opportunities to complete the academic part of the legal training, acceptance for vocational training in chambers for potential barristers will remain just as competitive.

The academic stage of training

Students with a 'qualifying' law degree – at least second-class honours – are exempt from the succeeding academic stage of training, the Common Professional Examination. Check carefully whether this applies to courses in which you are interested.

Students with a second-class honours degree in any other subject have to take a year's course leading to the Common Professional Examination. From September 1996 the coursework covers the foundations of legal knowledge – law of obligations (contract and tort), criminal law, equity and law of trusts, European Union law, property law and public law. Students also cover an extra legal study which will vary depending on the institution and will need to show expertise in legal research. Graduates with law degrees which are not listed as 'qualifying', because they do not cover all subjects, will need to take all or part of this course.

You must consider the fact that you are extremely unlikely to get an LEA award to follow such a postgraduate course. The first degree course – whether law or non-law – naturally attracts a mandatory grant.

The vocational stage of training

This usually consists of the one-year Bar Vocational Course, until recently only offered at the Inns of Court School of Law, but now available more widely at several institutions of higher education. The major part of this year of training is occupied with the acquisition of practical skills, and there is much less emphasis on the formal learning of facts. There are approximately three applicants for every place available, so there is now a selection procedure. Applications are made through a centralised clearing system known as CACH: contact the Education and Training Department of the Bar Council for details.

Pupillage

At present, all students who complete the Bar Vocational Course are eligible to be called to the Bar and may describe themselves as barristers. The newly trained barrister then has to do a year's pupillage with a barrister who is an approved pupil master, before starting practice on his or her own. This is when pupils really put their learning into practice. The second six months can be spent with an employer other than chambers.

Pupillage can be difficult to find, but the Bar Council will help to match pupils to vacancies via the Pupillage Applications Clearing House (PACH). It also publishes a handbook of chambers, pupillages and awards. In recent years, payment to pupils has become the norm – previously at least the first six months of pupillage was unpaid.

There has been a considerable expansion in the total number of barristers (there are about 7000 now, 22% of whom are women), but it is a very competitive and demanding profession, and intending barristers should consider other areas of the law for an alternative occupation, should they not be successful.

Qualifying as a barrister is a lengthy and expensive business. Living in London is very costly, and in addition to ordinary living expenses, buying a wig and gown and dining at your Inn of Court, fees for the vocational course will be incurred. Very few local education authorities give discretionary grants for these courses – check the position locally. There are some awards available from the Inns of Court and a few through the Council of Legal Education.

PROSPECTS

Finding a tenancy within chambers after completing the pupillage period can be difficult. Many barristers decide to use their qualifications away from the Bar at this stage. For a few years after qualifying, a barrister's earnings are likely to be very low – although the longer-term prospects can be excellent for successful barristers. Besides general practice, there are various areas in which a barrister can specialise, from family law to international practice.

Queen's Counsel

About one in ten practising barristers is a Queen's Counsel or senior barrister. This is an appointment granted by the Queen on the recommendation of the Lord Chancellor. Barristers who are not QCs are known as 'juniors' – whatever their age. A successful junior, after perhaps 15 years in practice, may apply to the Lord Chancellor to 'take silk'. QCs are known as 'silks' because they are entitled to wear a silk gown instead of a stuff one. By no means all applications are granted, but once appointed a QC a barrister starts on a new stage in his or her career. QCs are instructed only in the more important cases, concentrating on the legal problems involved and on advocacy. Procedural matters will be handled by a junior, whom the QC leads. Most circuit and High Court judges are appointed from the ranks of QCs.

Experienced barristers can also be appointed as a **recorder** (part-time judge) or a **stipendiary magistrate** (see later section).

Just THE JOB

BARRISTERS' CLERK

> Barristers' clerks are responsible for the administration of the set of chambers in which a group of barristers practise. Senior clerks are more like chambers' directors or chief executives. They are highly skilled managers. Most of the chambers are in London or the major cities. You can start work as a barristers' clerk with four good GCSEs or their equivalent.

What the work involves

Barristers' clerks have a good general knowledge of the law and its practice. They decide which briefs their barristers will accept from solicitors, and what fees will be charged. They organise the paperwork related to a particular brief, and make sure that everything is ready for a barrister to prepare their advocacy of a case. They also make sure that 'their' barristers appear in the right court at the right time. A great deal of responsibility goes with the job, as the clerk has to make many important decisions.

Junior clerks are trained on-the-job and do many different office tasks while they are learning the skills of office management and fee negotiation/administration. Their work can be physically demanding, transporting robes, papers and heavy books to and from court.

About three-quarters of the barristers practising at the Bar are based at the Inns of Court in London, so this is where most of the clerks work as well. There are also provincial chambers in some of the larger cities such as Birmingham and Manchester. Of the 300 or so senior clerks in England and Wales, only a small proportion are women, but more females are coming forward to do this work nowadays, and there is no lack of opportunity for them.

What it takes

Barristers' clerks must have a good deal of common sense and be able to use their initiative when called upon. They are:

- good communicators;
- good managers of people;
- reliable;
- able to cope under pressure;
- attentive to detail.

GETTING STARTED

Nearly all clerks start straight from school, and new entrants wishing to become fully qualified as a barristers' clerk normally need a minimum of four GCSEs at grade C, including English and mathematics, or the equivalent. Starting pay is competitive with other clerical jobs, and, on promotion to senior clerk (often by transferring to another set of chambers), payment is usually an agreed percentage of the fees earned by the barristers in the chambers. A successful senior clerk, or Clerk to Chambers, has an income which reflects their responsibilities and expertise. Some senior clerks earn more than the barristers they work for! The Institute of Barristers' Clerks can provide information on vacancies.

TRAINING

A junior can attend a two-year, part-time BTEC National

Certificate course, studying organisation, finance and management, law, marketing and chambers administration. After successfully completing the course, and after five years' experience in chambers, clerks can apply for qualified membership of the Institute of Barristers' Clerks.

just
THE JOB

LEGAL SECRETARY

Legal secretaries work in solicitors' offices, barristers' chambers, the law courts, banks, government departments and local authorities, the police, and any other organisations where their secretarial skills, combined with their legal knowledge, are required. You can train to do this work at college, or learn the legal side on-the-job if you already have secretarial qualifications.

The majority of the work of legal secretaries is like that of a personal assistant and includes typing or wordprocessing, taking shorthand, making appointments, etc. However, the legal secretary must also have a knowledge of legal terms and documents, plus some accounting skills.

The range of work varies from one employer to another, although being a legal secretary is mainly an office-bound job and there would rarely be an opportunity to go to court. Solicitors are the main employers of legal secretaries, especially outside London, and each solicitor generally specialises in one area of legal practice. Consequently, their secretaries will also.

The four main areas of specialisation are:

- **probate** – concerned with wills and inheritance;
- **conveyancing** – concerned with buying and selling property;
- **divorce and matrimonial matters**;
- **criminal cases**.

Larger firms may well specialise in one area of work alone. In London, for example, you will find firms specialising in finance and company law, or large-scale property transactions.

What it takes

- discretion – a great deal of your work is confidential, and you cannot go home and discuss cases with friends and neighbours;
- tact and sympathy – you may be dealing with clients who are distressed;
- accuracy, and a good command of English.

What the work involves

- **general secretarial work:** audio-typing, perhaps some shorthand, wordprocessing, filing, answering the phone and making appointments, dealing with general queries;
- **a basic knowledge of legal practice:** a legal secretary would never be required to give legal advice, but must understand the kind of problem a client has, and refer them to the appropriate person;
- **keeping accounts:** legal secretaries do not usually calculate fees, but are responsible for keeping records of work done and getting together the paperwork from which the fees will be worked out;
- **progress chasing:** legal secretaries need to keep a close check on how cases are progressing. This may involve ringing a magistrates' court, for example, to find out the result of a hearing about a client's maintenance payments.

Jobs are available at various levels ranging from receptionist/ general clerk to personal secretary/assistant posts. There may also be office manager posts in large firms. Your starting point would depend on your qualifications and whether you have any previous secretarial experience, either general or specialist.

Abroad, there are opportunities for legal secretaries to work in the European Courts of Justice, and in the overseas branches of British firms.

TRAINING

It is possible to enter legal secretarial work with a general secretarial qualification, and then to learn the specialised skills on-the-job. Would-be legal secretaries can study at college for National Vocational Qualifications (NVQs).

The Institute of Paralegal Training offers various part-time and full-time courses at colleges of further education, at two levels:

- Certificate level, which normally takes one year and requires a minimum of four GCSEs at grade C, including English language (or equivalent qualifications);
- Diploma level, which takes a further year and is of post-A level standard. This can be studied for part-time, but must be completed over three years.

Qualifications you gain on successfully completing any of these courses will, of course, also equip you for more general secretarial posts.

PROSPECTS

Legal secretaries with good qualifications may go on to qualify, through part-time study, as a legal executive, office manager or licensed conveyancer. A law degree, for anyone suitably qualified, would be the first step towards qualifying as a barrister or solicitor. Other possible careers are that of legal cashier and law costs draftsman:

- **Legal cashier** – for those interested in the financial and administrative side of a law practice. Student membership of the Institute of Legal Cashiers and Administrators is open to

anyone working in a solicitor's accounts department. No specific qualifications are required, but mathematics, English and business studies are useful subjects. The Institute's Diploma can be achieved through distance learning.
- **Law costs draftsmen** ensure that anyone who employs a firm of solicitors is charged a fair fee for the work done. They also help clients recover costs awarded to them as a result of a court case, or ensure that only reasonable costs are recovered by an opponent. Student membership of the Association of Law Costs Draftsmen is open to those in appropriate employment, and students work towards becoming a fellow of the Association through distance learning.

just THE JOB

LICENSED CONVEYANCER

> Transferring the ownership of a house, flat or piece of land from one person to another is known as *conveyancing*. A licensed conveyancer does the legal and administrative paperwork involved in the transfer, and 'searches' to see if any development is planned for the area which may affect the value or desirability of the property. Licensed conveyancers have the equivalent of a final professional law qualification, plus at least two years' practical experience.

Until 1987, the only people who could legally charge a fee for conveyancing were solicitors, although in fact the work was usually done by legal executives or conveyancing clerks working for a firm of solicitors. There is now a profession of licensed conveyancer, with its own regulatory body – the Council for Licensed Conveyancers.

Licensed conveyancers can be employed by other licensed conveyancers, solicitors, banks, building societies and local government 'in-house' conveyancing departments. Once qualified and holding a full licence, they may work either as a sole practitioner, or in a partnership approved by the Council for Licensed Conveyancers.

QUALIFICATIONS AND TRAINING

To register as a student of the Council for Licensed Conveyancers (or CLC), you need a minimum of four GCSEs at grade C from a list of approved subjects which include

English, or qualifications equivalent to these. Higher qualifications may be an advantage. Applications from people over the age of 25, without these qualifications but with relevant experience, are considered on individual merit. Students usually earn while they learn, supervised by a solicitor or licensed conveyancer.

Courses are offered by a number of further or higher education colleges, and normally involve part-time study. Full details are available from the Council of Licensed Conveyancers, which also provides a distance-learning course.

The examinations

The CLC examinations are in two parts. The foundation examination tests knowledge of the three modules: office practice and procedure, conveyancing, and land law and law of contract. The final examination covers conveyancing law and practice, landlord leases and tenancy agreements and accounting. In order to become fully qualified, students have to pass, or gain exemption from, all parts of the examinations. The exams are held twice yearly, and students must be registered before they intend to sit the exams.

Students also have to complete two years of supervised practical training. At least one year of this training must be undertaken after passing or being exempted from the examinations, before you receive your first licence. To become a licensed conveyancer, you also need to be at least 21 years old, and able to prove that you are a 'fit and proper person' to hold a licence. Even so, it is not until you have held three annual licences that you can set up as sole principal of a practice, following an interview and approval by the CLC.

PAY AND PROSPECTS

There are no formal salary scales: pay depends very much on the abilities of an individual and where he or she works. A good conveyancer should be able to command a salary comparable to other professions.

Students of the CLC can become associates of the Society of Licensed Conveyancers, the representative body, which publishes a professional journal and can offer assistance and advice to students and newly qualified conveyancers.

just THE JOB

WORKING IN COURT SERVICES

> There are various jobs in courts of law which help to keep the courts running smoothly. Some require legal training, but there are other opportunities involving clerical and financial work. The qualifications needed range from none, through a few GCSEs at grade C, or their equivalent, to a degree or postgraduate diploma.

Types of courts

Magistrates' courts are the first level in the court system in England and Wales. In these courts, found in the centre of many larger towns, a wide range of cases are presented for trial. For over 600 years, these courts have existed to deal with any local disputes and to restrain law-breakers. Magistrates' courts also deal with care orders and local administrative matters, like licences for betting shops and sellers of alcohol.

The cases heard in a magistrates' court range from minor offences, such as parking infringements, up to extremely serious crimes of violence and theft. All criminal cases start with a hearing in a magistrates' court and around 96% are concluded there. For serious crimes, the defendant can choose to be dealt with by the magistrates, or in a *Crown Court* before a judge and jury.

People read of the work in magistrates' courts through the reports in the local paper each week: *'man arrested after fight in pub'*, *'motorist drove at over 100mph through built-up area'*, and so on. Except for a number of stipendiary magistrates, the magistrates who conduct these cases are volunteers – not paid officials. The team of people keeping the processes of the law going are, however, paid officials. They are involved in collecting fines, preparing schedules of court business, dealing with correspondence and all sorts of other administrative matters. Throughout the country, over 9000 staff are engaged in the day-to-day running of magistrates' courts.

Above the magistrates' courts are the *Crown Courts*, which are part of the Lord Chancellor's Department of the Civil Service. The staff in the Crown Courts are civil servants and have the same pay and conditions as the rest of the Civil Service.

Minor civil cases involving small claims or property disputes are heard in *County Courts*; the more serious of these are taken to *High Court*.

Justices' clerk (clerk of court)

Justices' clerks advise magistrates on law and procedure, both in and out of court. Their duties may also include training magistrates and overseeing the general administration of the courts, as delegated by the justices' chief executive. To become a justices' clerk, you need to be a qualified solicitor or barrister of at least five years' standing. There are 220 clerks to the justices working throughout England and Wales.

Justices' chief executive

A justices' chief executive has the same qualifications as a justices' clerk. They are the principal advisers to the Magistrates' Courts Committee. Their precise role will vary from area to area, but, broadly speaking, they are in charge of the day-to-day administration of the courts in their committee's area.

Justices' clerk's assistant

This title covers quite a range of jobs which exist for considerable numbers of people in magistrates' courts. The work includes acting as court clerk, as a general administrative/clerical worker, as a keyboard operator dealing with all the typing work, or on the accounts side – dealing with financial matters.

Although the justices' chief executive has the overall responsibility for the administrative work of the courts in his/her committee's area, where several courts are in operation at once (and there can be up to twenty courts sitting in a purpose-built urban complex) experienced and qualified court clerks actually sit in court and advise magistrates on law, practice and procedure. The administrative work includes getting papers ready for cases, and doing all the necessary paperwork after the magistrates have made their decisions. There is also a considerable amount of work concerning such things as licence applications, adoptions and maintenance payments.

QUALIFICATIONS

There are no formal requirements for assistants, but in practice a minimum of three GCSEs at grade C, or equivalent qualifications, are required. Trainee court clerks need to have A levels or degree-level qualifications.

An assistant acting as a clerk of court must:

- be qualified either as a barrister or as a solicitor; *or*
- he or she must have a law degree with relevant experience; *or*
- have a diploma in magisterial law obtained through part-time study.

Basic training is given to all new entrants. There are more advanced courses in general administrative work, and also three-year part-time courses for intending court clerks, leading to a diploma in magisterial law. There are about 1800 court clerks working in some 700 courts throughout England and Wales.

There are occasional opportunities for solicitors to become justices' clerks by completing their training as articled clerks in the courts service.

Sean – court services officer

'I work in the pre-court section of the general office in a small town magistrates' court, and my job is essentially a backroom job. This doesn't mean, however, that things are quiet! It's a very busy office, with the public coming in with queries and requests for information, as well as all my routine work to do.

My job comes in two parts. Firstly, I have to prepare the summonses which are sent to people, telling them to appear in court. This means working with police and using a computer link to make sure that all the details are correct – dates and times, the offence committed and, of course, the name and address of the person involved. This, obviously, means I have to check everything I do very carefully.

I am also responsible, along with another colleague, for the court listings. We have to allocate times and dates for the hearing of the cases and make sure that all the paperwork is up to date and ready for the magistrates. This is a responsible job, but we get training and there is always someone to turn to if there is a problem I can't sort out.

I have to deal with the public as part of my job, as we often take a turn on the public enquiry desk where there is always some difficulty to be sorted out. Of course, we don't answer questions about the law, but we can help people to understand what the court procedures are all about. Other jobs in our office are concerned with all the business of what happens after a court case, recording the verdict, collecting fines, and so on.

I started my job straight from school with four GCSEs, but I'm doing some more studying and I'm hoping to move to more responsible jobs within court services.'

Financial staff

Even a small court may collect as much as £300,000 a year in fines and fees, and all this money must be very carefully accounted for. Increasingly, computerised accounting systems are being brought into use.

Usher

Ushers help to run the court. They call witnesses, check that defendants are present, instruct people on how to take the oath, and pass documents to the magistrate. There is no formal entry requirement for ushers, but it is very much a job for a mature person, not a school-leaver.

Court reporting

Conventional shorthand is no longer the method used to take down everything that is said during a trial and prepare a transcript when required. Approximately 50% of courts are now covered by tape-recording machines, requiring supervision to keep a detailed and accurate log of proceedings.

A computerised shorthand or *stenotyping* system is used in the other courts. Stenotyping requires a high standard of written English, as the transcriptions have to be a clear and accurate representation of each court case. An understanding of, and ability to spell, medical and technical terms in the field of engineering are also crucial. **Stenotypists** may be freelance or work for private reporting firms which provide a service to courts. There are a number of college courses available, some of which offer distance learning.

Coroner's officer

Coroners are appointed by a local authority to investigate unusual or unexpected deaths. They are assisted in their duties by **coroners' officers** – often ex-police officers – and **administrative officers**.

A coroner's officer needs to be mature and responsible, able to handle sensitive and sometimes tragic situations, and definitely not squeamish. The job may involve being on call at odd hours, and you are likely to need your own transport. The number and type of jobs available vary considerably between city and rural areas. (See later section.)

Bailiff

Courts contract out to companies which employ bailiffs to do the work of collecting fines and fees imposed through court judgments – a task formerly carried out by the police force. The job involves visiting those who default in payments imposed by sentencing magistrates or judges in order to work out a feasible routine for collection of money owed to the courts.

This work requires an articulate person with plenty of common sense and some maturity. No academic qualifications are needed for this job, which is open to both sexes.

JUST THE JOB

JUDGES & MAGISTRATES

> Judges and magistrates preside over courts, listen to cases and make judgments on the evidence they hear. They have the most responsible jobs within the legal system, having the power to deprive people of their liberty. Judges are qualified and experienced lawyers, whilst magistrates are normally mature volunteers who have usually followed a different profession.

Judges

Judges are almost always drawn from the ranks of experienced barristers who have become QCs (Queen's Counsels). To become a QC takes a number of years' experience, and a recommendation from the Lord Chancellor. After gaining considerable professional expertise at this level, QCs may eventually be appointed as one of the three hundred or so circuit judges in England and Wales and, ultimately, one of the much smaller, select band of about eighty High Court judges. These appointments are made by the Crown on the recommendation of the Lord Chancellor. Even higher positions, such as Lord Chief Justice, Judges of the Court of Appeal, and Master of the Rolls, are made on the recommendation of the Prime Minister.

A **recorder** is a part-time judge in the crown or county court, appointed to assist the circuit judges, and may be either a solicitor or a barrister by training, with about ten years' experience.

Magistrates

The situation for magistrates is rather different. There are two types of magistrate: the **stipendiary magistrate**, who, like a judge, is a paid official, and the **Justice of the Peace** (JP), who officiates in court on a voluntary basis.

- **Stipendiary magistrates** are usually found in urban areas (mainly London) and are trained lawyers, either solicitors or barristers, of at least seven years' experience. These appointments, as for judges, are made on the recommendation of the Lord Chancellor. There are about 80 stipendiary magistrates, as opposed to about 30,000 Justices of the Peace.
- **Justices of the Peace** are appointed by the Lord Chancellor, or the Chancellor of the Duchy of Lancaster, on the recommendation of an advisory committee of the local Commission of the Peace. They must be mature people of integrity and experience, at least in their late twenties, who are willing to give up a considerable amount of time to perform their duties. They receive some training before they start, and undergo further training during their first year as a JP. Refresher training continues throughout their time 'on the bench'.

All criminal cases start in the magistrates' court and around 96% of them are completed there. Magistrates usually sit as a group of three. They listen to the evidence, form a judgment and pass sentence in minor criminal cases. For a more serious crime, the defendant can choose whether to be dealt with by the magistrates or in a Crown Court in front of a judge and jury. Very serious crimes can only be heard before a Crown Court.

Magistrates' courts deal with more than just crime. In fact, they are part of the community, largely run by the local people. There are special courts to settle family disputes and make care orders for children. Magistrates also issue licenses for pubs, restaurants and betting shops. They are assisted in their job by paid court officials called **court clerks**, who are trained lawyers (usually solicitors).

Magistrates are unpaid, but can claim travel and subsistence expenses and an allowance for loss of earnings resulting from their duties.

Nominations for appointment as a magistrate can be made by local industry or community groups for consideration by the local advisory committee, or an individual can put his or her own name forward. You can get the address of the secretary of your local advisory committee from the Clerk's office at your nearest magistrates' court.

just THE JOB

REGISTRAR OF BIRTHS, MARRIAGES & DEATHS

The law requires that all births, marriages and deaths must be recorded, so every local authority must employ a Registrar to make these records. Births must be registered within six weeks, and deaths within five days. Marriages are recorded whether they take place in a church, a Register Office or elsewhere. Registrars need no specific academic qualifications.

What it takes

Registrars and Superintendent Registrars need to:

- have neat and legible handwriting;
- enjoy filling in forms and paying attention to detail;
- work strictly to regulations;
- be tactful and courteous;
- be able to deal with people in various emotional states;
- have a mature attitude.

This is not a job for a school-leaver; in fact, there is a lower age limit of 21.

Registrar

The Registrar is a record-keeper. He or she must be available during normal working hours to complete detailed forms required by law following a birth, a marriage or a death.

What the work involves

Obviously, in this work, you may be dealing with jubilant new parents at one moment, and recently bereaved people the next. You would need to be able to deal sympathetically with both these situations without becoming involved, as your job would be to gather and record the information precisely, both in handwriting and on computer. You would also need to be present at all marriages in Register Offices, or other approved premises, to complete the certificate, and at certain church weddings. Weekend working, therefore, is part of the job.

It is the duty of the Registrar to inform the Coroner if there appear to be any suspicious circumstances surrounding a death. Other duties include writing correspondence, keeping simple accounts and doing some statistical work.

In rural areas, the Registrar may be responsible for more than one district and therefore the Register Office may be open only at certain times during the week. In these circumstances, you would need to have your own means of transport.

QUALIFICATIONS

To be a Registrar, there is no educational qualification requirement, but, as you can imagine, great emphasis is placed on personal qualities, and job background. There are several *dis*qualifications for appointment. For example, if you are a minister of religion, doctor, midwife, undertaker or anyone involved in a business concerning life insurance, you cannot do this work.

Superintendent registrar

As well as a Registrar, each district also has a Superintendent Registrar. This person has overall responsibility for the Register Office and for all the records kept there, but does not write any of the records personally.

The Superintendent Registrar deals with all the legal preliminaries to a civil marriage in a Register Office or other approved premises. He or she must complete quite complex enquiries concerning the eligibility of a couple to marry – for example, have the couple been resident in the area for the required length of time, and are they legally entitled to marry (age, marital status)? By asking searching questions, the Superintendent Registrar can be satisfied that all the legal requirements are met.

He or she may be required to attend court if any of the information in the records is in dispute. The Superintendent Registrar also conducts the civil marriage ceremony, issues copy certificates on request, and helps people who are interested in investigating family histories.

A Superintendent Registrar needs to be an able organiser and must be confident when dealing with people. Imagine conducting a marriage ceremony in front of a hundred or more people! Training is on-the-job, provided by the Registration Training Officer.

just
THE
JOB

CORONERS & CORONERS' OFFICERS

> **Coroners** are experienced doctors or lawyers, who investigate any sudden or unexplained death. Every county has to appoint one or more coroners. They are independent of local and national Government, and act within certain laws and rules of procedure. **Coroners' officers** assist coroners with their legal duties.

Coroner

A coroner, or an appointed deputy, is on call at all times. A death would be reported to a coroner if the police, a doctor or the Registrar of Deaths was not happy about the cause. It might be that a death occurred:

- during an operation;
- in police cells;
- in a violent way;
- when there appeared to be no illness or obvious cause;
- when the deceased had not been seen by a doctor for two weeks or more.

In all these circumstances, the coroner would be called in. If the coroner is satisfied that death was due to natural causes, it can be properly registered, and burial or cremation can proceed.

If the coroner is *not* satisfied about the cause of death, an inquest has to be held. This is not a trial; it is an enquiry, and witnesses will be called. Its purpose is to establish the identity of the deceased, and the cause of death. In some circumstances, there is a jury to assist the coroner and it is the jury, rather than the coroner, that decides the verdict.

The inquest may help the family of the deceased to find out what happened. In the case of death from an accident at work, for example, an inquest may also help to prevent the same thing from happening again.

When the inquest shows that the death might be due to unlawful killing, the coroner must send the relevant papers to the Director of Public Prosecutions. The inquest does not try to find out *who* may have caused the death.

Although it is the job of coroners to seek the truth, to make sure that the inquest is run properly and no facts are covered up, they also realise that every death is a private tragedy to someone, and treat each one with dignity and sympathy. Personal information that has no relevance to the case will not be made public.

Who can become a coroner?
- a barrister or solicitor with at least five years' experience;
- a medical practitioner with at least five years' experience.

No councillor may become a coroner within their own county, or for six months after they cease to be a councillor.

Coroner's officer
A coroner is assisted by a coroner's officer, who is employed by the police and acts as a liaison between the police service, the coroner, the pathologist and relatives of the deceased. Because of their previous experience, coroners' officers are often ex-police officers.

What do they do?

Coroners' officers will be called upon to:

- attend sudden deaths to assess circumstances;
- obtain identification of the deceased, and previous medical history;
- collect statements and prepare reports;
- attend post-mortems;
- arrange and run inquests;
- carry out any other administrative tasks required by the coroner.

The job means being on call at odd hours. A coroner's officer needs to be mature and responsible, able to handle sensitive and sometimes tragic situations, and definitely not squeamish.

just THE JOB

THE PRISON SERVICE

> The main responsibility of the Prison Service is to see that people who have been convicted of an offence and sent to prison are kept securely in custody. However, an important objective of the service is to provide education, training and counselling to help prisoners find a life outside crime, and to deter them from returning to prison in the future. Prison officers do not need particular academic qualifications.

Some prisons are run by private companies, and their employment and training policies will differ from those described in this section, which describes careers within HM Prison Service.

Various kinds of custody are provided to suit offenders of different types and ages. Prison Service officers can work in:

- remand centres where people await trial or sentence;
- local prisons and training prisons (both 'open' and 'closed');
- young offenders' institutions.

The vast majority of the prison population is male, but there are jobs in the Prison Service for both men and women, who do not necessarily work with prisoners of their own sex.

The Prison Service includes governors (five grades), principal and senior officers, prison officers, psychologists, educationists, specialist trades officers, hospital officers, catering officers and physical education instructors. In England and Wales, approximately 40,000 people are employed in the Prison Service; over 27,000 of them are prison officers.

This kind of employment can have a considerable effect on your non-working life. Promotion may mean moving to another establishment. Shift work and night-work are required, which can affect family and social life. Officers have to cover bank holidays too, including Christmas Day! They work a basic 39-hour week. The Prison Service usually provides social facilities.

Prison officer

Prison officers come from all walks of life, and need:

- a balanced and mature attitude;
- patience and tact;
- common sense;
- a strong character, but with some flexibility;
- an ability to cope with a challenge;
- a sense of humour;
- a desire to make a firm commitment to the care of those in custody;
- a readiness to treat all prisoners equally.

TRAINING

Applications are made to individual prison establishments. Vacancies are usually advertised in Jobcentres. To train as a prison officer, you should be between the ages of 20 and $49\frac{1}{2}$ years. There is a minimum height requirement, but this is under review. There are no formal educational requirements.

Applicants take an aptitude test and have a medical examination and a selection interview. Personality is very important – there is a great deal of skill in establishing relationships with prisoners whilst still maintaining discipline. Some prisoners can be extremely difficult and potentially violent people.

When appointed, officers do eleven weeks' basic training, including nine weeks' residential training at one of the training schools (Wakefield or Newbold Revel). There are possibilities of promotion for capable officers to senior and principal prison officer grades and even to governor grades, subject to passing exams and promotion boards after a few years' experience.

Specialist training is available – for caterers, hospital officers and physical education instructors (with or without previous experience), and dog-handlers.

National Vocational Qualifications can now be achieved by employed prison officers. The scheme is open to all prison workers in the UK – about 60,000 at present – whose work can be assessed on-the-job to NVQ level 1, and to NVQ levels 2 and 3 where qualified assessors are presently available. As the qualification scheme is introduced to the service, the focus is on training and assessing new recruits and those who have only lately joined the prison service.

Governor

Governors are the managers in the Prison Service. The exact duties vary from one establishment to another, though staff and resource management are part of the work in all posts. It is a job for someone with qualities of leadership, a keen interest in social problems, organising skills and the ability to exercise authority, when needed, with firmness and sensitivity. Not only does the amount of responsibility increase as governors work their way to the top, but the hours of work are likely to increase too!

There are prospects for promotion even beyond being in charge of a major prison. Experienced governors can be promoted to area managers and operational directors.

At present, there are five grades of governor, but grades and promotion structures are under review.

Accelerated Promotion Scheme

Applicants with a degree or equivalent qualification (such as a BTEC HND) can apply to be accepted in the annual autumn intake for the accelerated promotion scheme, which is a direct route to governor status. Although graduates can be from any discipline, those with psychology, sociology or criminology backgrounds may be preferred. There is a tough three-day interview and a medical exam, after which the successful applicants start at the bottom of the ladder as basic grade prison officers, with the same practical training.

Someone on this scheme is likely to reach principal officer grade in four years, at which level they attend a six-month management training course, to reach lowest governor grade after five years. This is a considerably faster route to promotion than other officers can expect to follow.

Prison officers without degree-level qualifications can apply for the accelerated promotion scheme, if they show particular potential.

Specialist jobs

To work as a prison psychologist, trades officer or instructor (e.g. building trades, engineering), you should first do the required training and obtain the normal qualifications needed for the job concerned. There are also openings for qualified nurses. Teaching staff in prisons are often employed, not by the Prison Service, but by a local FE college or private training organisation. Religious workers of all denominations and faiths may work part-time or full-time within prison chaplaincies.

just THE JOB

SECURITY WORK

> Security work involves the protection of property, goods, money and people. It's a large and growing industry. Some organisations, such as retail stores, banks, local government and airports employ their own security staff; others rely on outside agencies. There are no formal academic requirements, but, obviously, personal qualities are very important.

Site security used to mean an elderly night-watchman huddled beside a burning brazier. A modern security officer is more likely to be wearing a uniform and patrolling with a two-way radio, or sitting in a control centre, keeping an eye on the instrument panel of an electronic burglar alarm or operating a closed-circuit television system.

The need for armoured vans and helmets in certain types of security work shows that the work can be dangerous, though this element should not be exaggerated. You should also note that there is a lot of work to be done at night and at weekends, which could affect family and social life, and that you may be working alone. Some part-time opportunities occur.

What the work involves

Industrial security guards are involved in patrolling factories, office blocks, warehouses and building sites at night, providing the first line of defence against fire, flood and robbery. They use two-way radios, electronic detection devices and smoke and heat detectors, so that one person can guard quite a large area. They are backed up by a control station which is equipped to give them immediate help in an emergency.

Mobile security guards do the same kind of work, but drive round in a van checking a number of factories, shops, offices and building sites during the night.

Drivers and crew in armoured security vans operate by day or night. Their duties include the safe transit of precious metals or jewellery, and carrying cash for banks, wages for large firms or the daily takings of supermarkets and other large retail companies. Drivers and couriers are required to provide a fast and dependable delivery service for packages and documents. Dog-handlers are needed in certain situations, and may be specialists or security guards trained in working with dogs.

We have all become familiar with armoured vans with uniformed and helmeted crews, but not all security operations are so noticeable. Security staff may mingle with the public whilst guarding important individuals, keeping a look out for trouble. Store detectives mix with shoppers, watching out both for shoplifters and for shop staff who are 'on the fiddle'. They may be employed by a store group or by a security firm. The job may involve moving around from shop to shop, so that their faces do not become familiar.

There are a few private prisons staffed and run by security firms. The work would be similar to that of a prison officer, but terms and conditions may be different. It could also include escorting remand prisoners to and from court appearances.

ENTRY AND TRAINING

Security work is not suitable for school-leavers, and there are often lower and upper age limits – frequently 18–55, though sometimes outside these. Many firms prefer ex-police, prison service, fire service or armed services personnel, or mature people with other suitable backgrounds. A driving licence is often required. Both men and women can undertake security work.

Obviously, one of the essentials for security work is to have a good character. Employers look for reliability, honesty, and common sense. You have to be physically fit. No particular educational qualifications are needed, though you will have to be able to write reports, to deal with complex security systems, and to comprehend the legal aspects of the work (how far a security worker can go, in preventing a theft, for instance).

Many companies have well-organised training schemes, on which trainees learn about their duties, as well as the legal aspects of the job. A number of organisations, including the Security Industry Training Organisation (SITO), run courses leading to NVQ level 2 in Security Guarding. There is also a security technician's modern apprenticeship for young people, leading to NVQ level 3. Some firms only offer on-the-job training with an experienced employee.

SITO also offers a range of distance-learning packages supported by SITO/City & Guilds examinations. These include Professional Guard Parts I and II, Cash in Transit, Retail Store Detection and Aviation. You can also study by distance learning, through the International Professional Security Association (IPSA), for the examinations of the International Institute of Security at levels 1 and 2 in Security Management, accredited jointly with City & Guilds. IPSA also offers short courses in specific aspects of the work. There are a few part-time courses in colleges of FE.

FINDING A JOB

There are many security firms, and addresses may be found in the *Yellow Pages*. The most reputable firms are registered with the trade organisation IPSA (see Further Information section). Jobs may be advertised in the local press and at Jobcentres.

just THE JOB

POLICE WORK

> Police officers are responsible for maintaining law and order, and ensuring that the public and their property are protected. They investigate crimes and are very much concerned with crime prevention. Everyone starts by serving as a police officer on the beat. Entry is possible from GCSE to graduate level.
>
> You have to remember it's not all out-and-about work. Police officers spend a lot of their time on paper work – usually the most unpopular part of the job!

County constabularies and other police forces

- Everyone joins the police as a **constable**, no matter what qualifications they have. It's essential to gain a thorough experience of basic police work before you can go on to higher-level posts and to manage others.
- You must be either a British subject, a Commonwealth citizen whose stay in the UK is not subject to restrictions, or a citizen of the Irish Republic.
- You need to be $18\frac{1}{2}$ to join a police force. The upper age limit used to be 30, but many forces are now looking for mature recruits, up to age 48 in some cases. A good educational standard is required and there is an entry test. It would be unusual for a young person with no exam passes at all to be selected. Graduates are able to apply for the Accelerated Promotion Scheme for Graduates (APSG), as well as for

normal entry. All entrants on the APSG complete the basic two-year probationary period.

- The police no longer have any height requirements. Health and physique must be good, and eyesight and health tests of a high standard must also be passed. Nearly all forces accept applicants who wear glasses or contact lenses, providing unaided vision reaches a certain standard. Normal colour vision is also required. Check with individual forces.
- Character and personality are of great importance in being selected for the police. Qualities like maturity, common sense, discretion and the ability to keep calm in emergencies are the kinds of thing recruiters look for.
- Very few forces recruit cadets aged between 16 and 18.

The first two years

After a few weeks' induction with your local force, you go on an intensive training course of ten weeks at a police training centre. This gives you a basic grounding in law and police procedures, and shows you and your instructors how well suited for police work you are.

You learn about the legal system, court procedures, how to deal effectively with people in all types of situations, and topics such as traffic regulations/offences, arrests, missing persons, etc. There are theory and practical classes, and physical exercise periods for drill, lifesaving, self-defence and sport.

After this initial training, you return to your force for five weeks to put what you have learned into practice, under the guidance of a tutor constable. This is followed by a further period at the police training centre to learn the more complex elements of policing. You then, again, return to your force and really begin to learn how things operate in practice – on the beat, on patrol in the Panda cars, in court, and so on. You attend occasional training sessions, and also go on attachments to specialist departments – CID, traffic, fingerprints, criminal records, dog-handling.

Training with the Metropolitan and Scottish police forces is slightly different.

It's a highly responsible job

In police work you have to take many decisions yourself on a day-to-day basis. You have to use your judgement to deal with difficult situations. All officers have to cope not only with physical danger and abuse, hooliganism, etc, but also with situations like breaking bad news to victims' relatives, being first on the scene after bad traffic accidents and suicides, or dealing with very unpleasant crimes. At the other end of the scale there are the jobs like giving talks about your work to children, crossing patrol duty, routine court appearances, etc.

You'll have to get used to the hours. Police officers work an eight-hour day, usually on a shift system which includes nights and weekends. You have rest days at the rate of two a week, but not always two consecutive days. You get about one weekend off each month, and, of course, take your turn at bank holiday duties and so on.

PROSPECTS FOR PROMOTION

Promotion and specialisation depend solely on your own abilities and the recommendations of your senior officers. Recent figures showed there were about 92,900 police constables, 19,800 sergeants, 6800 inspectors, 2300 chief inspectors, and 1500 superintendents. These figures may give you an idea of the chances of gaining promotion.

Most constables who gain promotion make fairly slow progress up the ladder. The first step is to **sergeant**, which you could achieve within five years or so, and then **inspector** – which

> ### Julie – police inspector
>
> I went to university after I left school, and then I joined the police force. All police start as constables and do two years' training, no matter how many exams they have passed. The training includes some time spent at college, learning about the law and policing, and some time at a station, actually doing the job – meeting people and dealing with day-to-day problems as they crop up. It's good, because you learn thoroughly what policing is all about. But I wanted promotion. I wanted a challenge and I wanted to be the one making the decisions. This is where my extra time spent at university has helped me, because I was able to take my sergeant's exam after three years and have just become an inspector four years later.
>
> Now I'm like the manager of a big team, leading two sergeants and twenty constables at my station. On day shifts, most of my time is spent on planning and dealing with problems as they come up. I don't get much chance to go out, except when there is a major incident like a murder on my "patch". Then, as inspector, I lead the enquiry. Night shifts are my chance to spend some time with the team, and go out on the beat with them as much as I can.
>
> One of the good things about this job is the chance for promotion, as long as you are prepared to keep studying and taking exams.

you could attain after about nine years' total service. At any rank you can apply for transfer to another force, and experience in other forces is essential for the more senior postings.

Recruits with outstanding potential may be selected for the

Accelerated Promotion Course at Bramshill Staff College, possibly gaining promotion to **inspector** in only five years. This is the same course as taken by the high-flying graduate entry. The total intake is not more than 80 per year, so competition for places is fierce. Another opportunity is for officers to be sponsored on full salary to take a degree course in an appropriate subject, either by Bramshill or through their own force.

Opportunities for specialisation

After successful completion of the two-year probationary period, you can apply to join one of the specialist branches. There are, however, relatively few openings each year so the competition is quite strong.

Traffic police

The traffic department is concerned with every aspect of safety on the roads. The work can include checking that vehicles conform to safety requirements, escorting abnormal loads, dealing with traffic accidents, keeping traffic flowing through busy towns, dealing with motoring offences like speeding and drunken driving and coping with victims of accidents. Drivers of traffic cars and motorcycles (different from the Panda cars which belong to the local police stations) go out on patrol and deal with incidents needing police presence. They are trained to drive safely at high speeds, and members of the traffic department have to learn about all aspects of traffic work.

The CID (Criminal Investigation Department)

A lot of CID work is painstaking routine. Most of the crimes dealt with by the CID are the more serious ones of burglaries and robberies, serious assaults and murders, fraud, rape, sexual offences, etc. There are specialist sections such as the Fraud Squad, Drugs Squad, Regional Crime Squad – these are only open to experienced detectives. CID hours are long and irregular.

Crime prevention officers

CPOs are specially trained to advise on security systems for industry, community crime prevention, house and vehicle security and personal protection for members of the public. All forces have crime prevention branches and the larger ones have local crime prevention officers.

Dog-handlers

Police dogs and their handlers are both very highly trained – you may well have seen them giving displays of their skills. Handlers look after their own dogs, which live with them as part of the family. Handler and dog become an inseparable team – the dog is normally taught to answer only to its own handler's orders.

Mounted police

This is a very popular branch with a long waiting list, but you don't have to be an experienced rider to apply. You should be prepared to stay a good few years even though promotion is limited. As well as crowd control, where horses are invaluable, mounted police are used in searches over moorland, etc, and go out on regular patrols. Besides riding, you see to the general care of your horse and equipment.

River police

Patrol boats police the river, dealing with craft in difficulties, thefts from boats, warehouses, etc, rescue work, smuggling, keeping an eye on riverside and harbour buildings and landing places, checking things like oil pollution, and helping in pursuits when called in by the CID or uniformed branch. Competition is strong for the places available and you have to be a very good swimmer. Competence with boats is certainly an asset.

Underwater units

Very few opportunities occur for officers to train as divers, who search for missing persons, abandoned stolen property, murder weapons, etc. The work can be dangerous and unpleasant, and calls for someone extremely fit.

Armed police

Some police are trained to use firearms when needed as a last resort and police marksmen are experienced officers. Training is rigorous and there are very frequent refresher courses.

British Transport Police

The BTP polices the railways, stations and other premises used by the British Railways Board and London Transport, including some Channel Tunnel operations. Ranks and rates of pay are as for the ordinary police. There are uniformed officers, detectives, specialist officers to deal with major thefts, raids, etc, and dog-handlers. Entry requirements and training are similar to those for the ordinary police. There is no cadet scheme.

Other specialisations include:

- **Special Branch** – working on surveillance duties (you may require CID experience);
- **Mobile Police Support Units** (brought in at short notice to assist where large numbers of officers are needed for searches, etc);
- **Juvenile Bureaux**;
- **Operations Room** work – where 999 calls are received and information is fed out to patrol cars, local stations, mobile units, etc;
- **Child Protection Units**, working closely with social services departments.

Other police work
Special Constables
'Specials' work full-time in other jobs but give up some of their free time to be part-time police officers, as a way of helping the community, for anything from a few hours a week to a few hours a month. Specials have the same responsibilities and authority as a regular officer. You are unpaid but receive expenses and a free uniform and equipment.

Ministry of Defence police
The Force polices the interests of the Ministry of Defence throughout the United Kingdom. Its role is similar to that of other police forces, with the added responsibility of physical security. Pay and conditions are linked to Home Office Police Forces. Specialised departments include CID, Dogs and Water Police.

Civilian jobs with the police
In all police stations and headquarters, the work of the police is supported by civilians, who provide a variety of clerical and technical support services. Many of the jobs described are available in all forces, but others may only exist in large forces such as the Metropolitan Police. Some forces are trying to hand over more work to civilian staff, leaving the police officers more time to do what only they can do. For most of these jobs, uniforms are not worn – but you are still very much a part of the police force you work for.

Administration, clerical and computing jobs
Police headquarters and police stations have similar administrative needs to any other organisation, in departments such as personnel, finance and computing, as well as in more specialist areas. Police and civilian clerks work together. Their duties are very similar, except that the police clerk is able to offer specialist

advice on points of law or police procedures. The work may include filing, gathering statistics, dealing with correspondence, reception and telephone enquiries. A clerk could be working with the CID, in the traffic section, or preparing the police case for a prosecution.

A general education to a good GCSE standard is the usual entry requirement. A level and graduate entrants may be taken on as administrative officers or management trainees. Computer skills may be of particular use, as computer databases are such an important part of police work today.

Traffic warden

A well as reporting parking offences, traffic wardens may direct traffic and deal with other traffic offences. Applicants must be over 18 years of age (older in some forces). There is no minimum height requirement, but wardens must have British nationality and good eyesight (spectacles and contact lenses permitted). There are written tests in English and mathematics, and a medical examination. After a short initial training course, further training is provided 'on-the-job'.

Law clerk

Law clerks are employed by the Home Office in Crown Courts, helping to prepare police cases for trials. They attend court and work closely with barristers and police officers. You must be over 18, with five GCSEs at grade C, including English, or equivalent qualifications.

Scene of crime officer

The Police employ civilians to work alongside police officers, carrying out examinations at the scene of a crime. They may also be called upon to give evidence in court. A good general education, with a scientific bias, is required. The job usually involves working shifts.

just THE JOB

PRIVATE DETECTIVE WORK

> Private detective work is usually concerned with civil legal business, and rarely with crime. Most of the investigations are routine, and are made on behalf of commercial organisations and solicitors. No formal educational requirements are necessary, although, of course, personal qualities are important.

What the work involves

A firm that needs to trace a bad debtor, or control thefts from its warehouses, might call in an investigation agency. At a more sophisticated level, it may be necessary for a high-tech firm to take precautions against industrial espionage or infringements of its patents' rights. A firm may want to investigate the genuineness or business status of a customer they are doubtful about. Solicitors working on a case for a client may need to have a writ served on someone they cannot contact, or find someone needed to provide evidence for the preparation of a case. Rather than spend time themselves they will turn to an agency for assistance.

As well as detective work, agencies are often involved in such activities as debt-collecting, supplying credit references, investigating insurance claims and acting as bailiffs.

The work requires a persistent and patient type of person, with good powers of observation and an ability to stand exposure to the seamier side of life. People often do not take kindly to having writs served on them or having their car repossessed. A good investigator will have a thorough knowledge of the law relevant to their field of work. Many agencies recruit staff who have had police experience of some sort, as, of course, this is very relevant.

In general, this is not a job for young people starting their first job, though the larger agencies need clerical and secretarial staff who might, with experience and maturity, move on to detective work.

QUALIFICATIONS AND TRAINING

There are no formal requirements for this kind of work, and most training is done on-the-job. However, the Institute of Professional Investigators now offers National Vocational Qualifications in Investigation at NVQ levels 3 and 4, which can be gained by people already in the industry, or by a distance-learning route. There are also postgraduate courses in Investigative Management at Loughborough University. The Association of British Investigators holds seminars for those people already working in the industry.

FINDING A JOB

At the present time, you do not have to be licensed in this country in order to practise as a private investigator. You will find addresses of private investigation agencies in the *Yellow Pages* directory or, alternatively, you could obtain a directory of agencies from the Association of British Investigators.

just THE JOB

FORENSIC SCIENCE

> The staff of forensic science laboratories work closely with the police on the scientific investigation of crime. Scientific staff analyse materials which might link the scene of a crime, a victim or a weapon with the criminal, or otherwise establish an essential element of the crime. Forensic scientists are graduates; their assistants have lower qualifications.

What the work involves

Forensic science involves the examination of such things as blood, human organs, tissues and body fluids, hairs and fibres from clothing, paints, glass, metals, firearms, bullets, tools, and explosives. Forensic scientists provide scientific support in the investigation of crimes against the person, crimes against property, and identification of drugs and poisons. They help police enquiries into explosions, suspicious fires, murders and motor vehicle accidents.

Forensic science labs have various sections, based on the main science disciplines of biology, chemistry and physical science, and including firearms and forgeries. Photography and information are provided as services for the other sections. The laboratory work consists mainly of chemical and biological techniques: chemical analyses, microscopy, blood testing, ultra-violet, infrared and X-ray spectroscopy, radiography, chromatography, emission, absorption and mass spectrometry.

Visits to scenes of crimes may be necessary to collect material for laboratory examination, though this is often done by police 'scene of crime' officers. Results are produced as a witness statement, or scientists may appear in court.

What it takes

This is not suitable work for anyone who is squeamish. Forensic scientists must be careful and methodical and pay a great deal of attention to accuracy and detail. Good colour vision is essential. Those responsible for work such as writing reports and presenting evidence must be good communicators, in both speech and writing.

EMPLOYMENT AND TRAINING

The Forensic Science Service has seven laboratories, at Aldermaston, Birmingham, London, Chepstow, Chorley, Wetherby and Huntingdon, each employing about 100 staff. There is also a laboratory in Northern Ireland and four police laboratories in Scotland. Entry is possible with various qualifications. The service employs a mixture of graduate forensic scientists, and laboratory technicians/assistants with GCSEs, A levels or BTEC qualifications. Advanced level GNVQ may also be acceptable. It is a relatively small service, and there is always a lot of competition for posts.

There are distinct job levels in the forensic science service. Once in the service, you can progress from one level to the next, if opportunities arise:

Assistant forensic scientist: a minimum of four GCSEs at grade C (or equivalent) is required. These must include English, and mathematics or a relevant science. Applicants with A levels, BTEC qualifications or Advanced GNVQ may be preferred. There are two salary grades at this level.

Forensic scientist: you need a good honours degree (or equivalent) in an appropriate science subject, usually chemistry or biology based. There are two salary grades to progress through.

Senior forensic scientist: besides a relevant degree, applicants need some postgraduate research or employment experience. Promotion to this level, which has three grades, is usually from within the service. It is possible to specialise in particular areas of forensic science.

Doctors and dentists: There are very limited opportunities to specialise in forensic medicine. The first step would be to qualify as a doctor or dentist in the normal way.

Specialist courses
There are postgraduate courses in various aspects of forensic science. Strathclyde and Bradford Universities offer first degree courses with forensic science as a key element. A few other degrees allow forensic science to be studied in conjunction with other subjects.

Adults: note that maturity and previous experience may mean that stated entry requirements can be relaxed.

FINDING JOB VACANCIES

Assistant forensic scientist vacancies are advertised in the local press and at Jobcentres. Scientific officer vacancies are advertised in the national and scientific press (e.g. *New Scientist*). Information about both levels can also be obtained from the Forensic Science Service Headquarters.

IMMIGRATION, CUSTOMS & EXCISE

> Many jobs in the Civil Service are for general entrants who can work in any department. This section gives information about specialist posts where particular knowledge or skills are required.

HM Immigration Service

The Immigration Service is part of the Immigration and Nationality Department of the Home Office. It is responsible for maintaining an effective immigration control at ports of entry to the UK, with the minimum of inconvenience to the travelling public.

Immigration officers are based at air and sea ports around the UK and the Channel Tunnel. They need a good knowledge of current affairs. They spend most of their time interviewing the public, so they need to be able to communicate clearly and persuasively, and have good listening and report-writing skills. Opportunities are nearly always in the south-east – especially Heathrow, Gatwick and Dover. Major ports are open all the year round, so officers work in shifts including nights, some weekends and bank holidays. There are also administrative officer and assistant posts in the Department's headquarters in Croydon.

Customs and Excise

Customs and Excise collects a third of central tax revenue. It's not all about searching bags at airports and ferry terminals!

Calculating the duties levied on imports is complex and time-consuming, with fast-changing rules and regulations. Since 1992, the freer movement of goods and people within the European Union has meant that fewer uniformed Customs and Excise officers are needed to seek out contraband, although they are still required to try to halt the movement of illegal drugs, animals and plants, armaments, etc. Many officers are involved in the work of collecting the excise duty on a range of goods produced in this country, including alcohol and tobacco. An executive officer may work full-time with a large distiller or tobacco company, ensuring that the correct excise duty is being paid. Betting duty also has to be collected, which involves officers in checking bookmakers' accounts. A major part of the department's work is the collection of VAT. Inspectors work out of the office most of the time, checking on businesses registered for VAT.

There are also many office-based administrative jobs. Officers need to understand and apply a range of complex rules, laws and accounting procedures and to be able to deal with people in difficult situations. Training is an important and continuing part of the job.

The bulk of Customs and Excise staff work in the network of regional offices, although about one-fifth of the workforce are based in headquarters in London, Liverpool, Manchester and Southend.

PATENTS & TRADE MARKS

> A patent is granted to a company or individual who claims to have invented a new product or process. The patent protects the invention from being copied for up to 20 years, whilst it is being made, used or sold by the originators. **Patent agents** present new inventions to **patent examiners** for investigation. Because most patents are for highly practical and technical devices, agents and examiners in the Patent Office are usually people who have gained a degree or equivalent in science, mathematics or engineering.

Anyone who wishes to patent an invention has to prepare a very clear description of it and then submit this to the Patent Office of the Department of Trade and Industry or the European Patent Office in Munich. Presenting an invention to be patented requires care and experience, so most inventors turn to a patent agent for assistance. The patent agent advises the inventor on the likelihood of the invention being original, and then presents the case to a patent examiner, receiving a fee from the inventor for carrying out this work.

Patent agent

In order to secure a patent, a full description of the invention needs to be filed with the Patent Office for examination, and most inventors use the skills of the patent agent to do this. All patent agents need a scientific or technical background, usually a science or engineering degree. They must also have knowledge of the law and a good command of English. Because patents are becoming increasingly international, foreign language skills are highly desirable.

Many patent agents work as technical assistants or partners in firms of patent agents. Others work in the patent departments of large firms or in government departments. London is the main centre for agency work. There are also opportunities in other large industrial centres, such as Birmingham, Manchester, Liverpool, Glasgow and, increasingly, where new industries are developing.

There is also the European Patent Office and other opportunities in overseas work. Many patent agents take not only British exams, but also those of the European Patent Institute. At present, most of the patent work in Europe falls to the British and German patent agents.

Carmen – patent agent

'A patent is a legal way of protecting someone's invention, so that no one else can make a fortune by passing it off as their own. When I talk about inventions perhaps you think of mad, white-haired inventors wearing six pairs of glasses with a new mousetrap they have just come up with. In fact, inventions are what is called 'intellectual property'. A formula for a new medicine is just as much an invention as a mousetrap, and can make an enormous amount of money for the inventor, who in this case will probably be an international drug company.

In my job, I have to be an expert on patent law, but I also need to be a scientist and a bit of a linguist. I have to produce a description of my client's invention, which is accurate enough to be lodged with the government patent office. To produce this description, I need a lot of

> specialised knowledge of the law, and I need scientific knowledge to understand what is new about this particular invention. Once the patent is lodged, for a period of time, the inventor is the only person who can gain from their work. This could mean selling the rights to the patent to a big firm who are in a position to manufacture it.
>
> I may also need to be able to look at French and German patents, to see if anything similar has been registered as a European patent. This only means reading these languages; I don't have to speak them fluently. This European-wide knowledge is very important. To have the best career opportunities, you need to have passed the European qualifying examinations.
>
> I started as a scientist, by doing a chemistry degree at university. When I started patent work, I had to do a lot more legal and practical training to become a fully registered patent agent. I work for a firm of patent agents and I hope that, eventually, I will become a partner in the firm.

QUALIFICATIONS AND TRAINING

This is mainly a graduate profession and a scientific, mathematics or engineering degree is usually necessary. The ability to write clearly and concisely is important. Trainees are called **technical assistants** and study for the examinations of the Chartered Institute of Patent Agents and, in most cases, the qualifications required by the European Patent Office.

The minimum period of study for a graduate is three years but this can often extend to four or five, as many candidates do not pass the examinations on the first attempt. Various part-time tuition is available. During training, the emphasis is on acquiring practical skills rather than purely academic knowledge and, before he or she can be registered as a patent agent, the trainee must have spent an appropriate amount of time working under the close supervision of a qualified patent agent. Despite the high qualifications necessary, competition for posts is quite stiff.

In the course of patent work, there is a lot of communication between British agents and their counterparts in other countries. A British patent agent is likely to file applications for patents in other countries, using foreign colleagues as intermediaries. To practise before the European Patent Office, you have to be on the list of European Patent Attorneys, for which you have to pass the European Qualifying Examination (EQE). An ability to read documents in French or German is necessary to achieve this, but the level of knowledge of languages required is not unduly high, and assistance is normally available for candidates taking this examination.

Graduates must spend at least three years in the profession (non-graduates, longer) before the European examination can be taken. There are a few postgraduate qualifications which grant exemption from some exams.

Patent examiner

Patent examiners are employed by the Patent Office, which is an executive agency of the Department of Trade and Industry in Newport, South Wales. About 1000 people are employed by the Patent Office, of whom 200 are examining staff. The annual number of vacancies (if any) is very small. Patent work is at the leading edge of any technology, and, to be able to understand an application, the examiner must have the appropriate skills and knowledge. Consequently, patent examiners usually work within specific subject areas.

All incoming patent applications have to be carefully dated, checked and studied. The examiner then starts a search, which usually takes about three months, making certain that the claimed invention is new and not simply a development of something which already exists. Details of the application are published once the search is complete, to allow anyone to study it. Further examination and report-writing is carried out, and it may take up to four years before all the legal and technical requirements are met and a patent is granted.

QUALIFICATIONS AND TRAINING

Applicants to patent examining should have a first or second class degree in physics, chemistry, mathematics or engineering, or an equivalent qualification. Training is given in the principles of patent law and, by working with experienced examiners, new entrants gradually acquire expertise in a particular field of invention. They must also become reasonably fluent in French and German.

Adults: it is possible for people qualified in law, engineering or science to move into patent work later in their career. This usually involves working in the patent department of an industrial firm rather than patent agency or examining.

Trade marks

This is a related area of work. Trade marks are registered in the same way as patents, though, unlike patents, a trade mark can be kept indefinitely. The larger patent agencies have trade marks sections. The work is not scientific or technical but more concerned with legal matters, as there is a large body of legal judgments and cases about trade marks.

just THE JOB

COMPANY SECRETARY

> Company secretaries, also known as chartered secretaries, are highly qualified professional administrators and managers. They are a company's legal representative, involved with the organisation of company business. Company secretaries have financial and legal skills, as well as administrative capabilities. There are professional qualifications for company secretaries, and many have degrees.

The company secretary acts as the link between the shareholders who finance a company and the directors who run it. He or she is a senior member of the firm's administrative staff, right at the centre of the business, entrusted with much confidential information. This is the origin of the name *secretary* – a keeper of secrets.

What the work involves

A company secretary's work covers three main elements:

- **Legal work** – the company secretary makes sure that the company works within the law, keeping records required by law and obeying company legislation. It is essential to have a complete understanding of relevant law, although lawyers would be used when necessary.
- **Assisting the board of directors** – a company is run by its board of directors, who meet regularly to plan business activities, review progress, examine policy, etc. Meetings of the board are organised by the company secretary, who arranges the agenda (the list of business items to be discussed at the

meeting) and gives the directors the background information on which to base their decisions. In order to present complex information to the directors in a understandable form, the company secretary must understand all aspects of a business. An analytical mind is needed for this, as well as good communication skills.

The company secretary also organises the Annual General Meeting of company shareholders and must ensure that the AGM is run in such a way that it complies with the Companies Acts. Often the company secretary is called upon to advise the chairperson on legal points.

- **General administration** – this varies greatly from one company to another, according to the firm's requirements. Job descriptions could include other specialist functions, such as personnel, data processing, pensions management.

George – company secretary

"You may think that, with a job title like mine, I spend my working days taking shorthand and typing letters. In fact, I don't know any shorthand, and I can only type very slowly with two fingers!

My job is quite different. I am a legal adviser and professional administrator in the company I work for. I am right in the centre, acting as a link between the shareholders (those who own the company) and the directors (those who run the company). I advise on all legal and administrative matters, making sure that the company operates within the law, and follows all the correct procedures.

I also have to attend high-powered board meetings, just in case my legal knowledge is required. It's exciting, being involved with those who make the big decisions. I enjoy being an important member of the team.

> *I joined this company after university, and have a bit more studying to do part-time for professional qualifications. You don't have to have a degree though – you can do all your studying on a part-time basis while you are working.*

QUALIFYING AS A COMPANY SECRETARY

The professional body which offers qualifications for intending company secretaries is the Institute of Chartered Secretaries and Administrators (ICSA). However, by no means all people who take ICSA qualifications end up as company secretaries – it provides a very useful professional qualification for many other kinds of administrative work too.

The ICSA syllabus covers four core areas fundamental to the different roles of administrators and managers. These are law, finance, management and information systems. The qualification consists of seventeen modules (or subjects) separated into three programmes – the Foundation Programme, the Pre-Professional Programme and the Professional Programme. The Foundation and Pre-Professional stages provide a broad-based introduction to the business environment.

The Professional Programme covers the essential areas of corporate law, finance, regulation and tax.

Access to the ICSA courses is dependent upon the qualifications you already hold. You may be eligible to gain exemption from certain ICSA studies. For example, holders of BTEC Higher National awards, legal executives, accounting technicians and others can gain exemption from the Foundation Programme and possibly parts of the Pre-Professional Programme. Holders of degrees and certain professional qualifications can apply directly for the Professional Programme.

The ICSA also operates an 'Open Access' policy for those who are 17 years old or over, but who do not hold any qualifications. Open Access entrants start the ICSA examination studies on the Foundation Programme.

TRAINING ROUTES

- Full-time ICSA courses are available at some centres, or there are certain degree and postgraduate courses which allow maximum credit towards the ICSA qualification.

- You can enter employment and study part-time towards the ICSA qualification. On average, it takes about four years to complete all seventeen modules, and there is a limit of five years allowed for completion. Study can be at a college or by correspondence.
- ICSA Certificate courses are offered in a range of administrative and managerial subject areas, e.g. Company Secretarial and Share Registration Practice. These courses equate to the Foundation Programme of the ICSA Professional qualification, and no formal entry requirements are specified.
- Some postgraduate courses result in a postgraduate qualification plus ICSA membership. Others give exemption from most of the ICSA Programme.

Whichever training route you take, several years of relevant work experience are also required before you can be elected to full membership of the ICSA.

ICSA examinations are relevant for work in central and local government administration, administrative or management posts in industry or commerce, voluntary or other public service organisations and consultancies – in fact, wherever a broad administrative qualification would be useful. ICSA qualifications are recognised throughout the European Union.

FOR FURTHER INFORMATION

GENERAL

The Law Society of England and Wales – 227–228 Strand, London WC2R 1BA. Tel: 0171 242 1222.

The Law Society of Northern Ireland – Law Society House, 90–106 Victoria Street, Belfast BT1 3JZ. Tel: 01232 231614.

The Law Society of Scotland – Rutland Exchange, Box ED1, 26 Drumsheugh Gardens, Edinburgh EH3 7YR. Tel: 0131 226 7411.

Careers in the Law, published by Kogan Page.

CRAC Degree Course Guide – Law, available in most school, college or careers centre libraries.

Getting into Law, published by Trotman.

The Legal Profession 199– is an annual Ivanhoe guide from Cambridge Market Intelligence: order from CMI, London House, Parkgate Road, London SW11 4MQ. Tel: 0171 924 7117.

The Legal Profession is an AGCAS booklet, obtainable from CSU, Despatch Department, Armstrong House, Oxford Road, Manchester M1 7ED. Tel: 0161 236 9816, ext 250/251. It may also be available for consultation in your careers centre library.

OPPORTUNITIES OUTSIDE PRIVATE PRACTICE

Crown Prosecution Service – Headquarters, Personnel Branch 2, 50 Ludgate Hill, London EC4M 7EX. Tel: 0171 273 8309. Publishes *The Case for the Crown Prosecution Service*, detailing opportunities for lawyers and trainee lawyers in the service.

Fast-stream and European Staffing Division (Civil Service) – Cabinet Office, OPSS, Horse Guards Road, London SW1P 3AL. Tel: 0171 270 6293.

Lawyers in Local Government – c/o Jacqui Dixon, Portsmouth City Council, Civic Offices, Guildhall Square, Portsmouth PO1 2QJ. Tel: 01705 834865.

The Law Society – Careers Office, 227–228 Strand, London WC2R 1BA. Tel: 0171 242 1222. Publishes booklets entitled *Local Authorities* and *Commerce and Industry*.

SOLICITORS

The Law Society of England and Wales – Careers and Recruitment Office, 227–228 Strand, London WC2R 1BA. Tel: 0171 242 1222. The Law Society publishes *Becoming a Solicitor* as well as booklets on a career as a solicitor in industry and commerce, in local authority departments, and in private practice.

BARRISTERS

Council of Legal Education – 39 Eagle Street, London WC1R 4AJ. Tel: 0171 404 5787. Organises training for the Bar.

General Council of the Bar of England and Wales (Bar Council) – 3 Bedford Row, London WC1R 4DB. Tel: 0171 242 0082. Provides information and guidance about the training and work of an intending barrister in *The Bar into the 1990s*.

BARRISTERS' CLERKS

Institute of Barristers' Clerks – 4A Essex Court, Temple, London EC4Y 9AJ. Tel: 0171 353 2699.

LEGAL EXECUTIVES & LEGAL SECRETARIES

Association of Law Costs Draftsmen – Church Cottage, Church Lane, Stuston, Diss, Norfolk IP21 4AG.

Institute of Legal Cashiers and Administrators – 2nd Floor, 146–148 Eltham Hill, Eltham, London SE9 5DX. Tel: 0181 294 2887.

Institute of Legal Executives (ILEX) – Kempston Manor, Kempston, Bedford MK42 7AB. Tel: 01234 841000.

Institute of Legal Secretaries – ILS House, 76 Denison Road, London SW19 2DH. Tel: 0181 540 0512.

Institute of Paralegal Training and Association of Legal Secretaries – The Mill, Clymping, Littlehampton, West Sussex BN17 5RN. Tel: 01903 714276. The Institute can provide a list of courses for their examinations.

National Association of Paralegals – Castle Circus House, Union Street, Castle Circus, Torquay, Devon TQ2 5QB. Tel: 01803 299956.

LICENSED CONVEYANCERS

Council for Licensed Conveyancers – 16 Glebe Road, Chelmsford, Essex CM1 1QG. Tel: 01245 349599. This is the regulatory body responsible for licensing conveyancers, maintaining professional standards and for setting the examinations. The CLC publishes a useful booklet which gives details of current training requirements.

Society of Licensed Conveyancers – 55 Church Road, Croydon, CR9 1PB. Tel: 0181 681 1001. This is the body responsible for promoting the profession.

WORKING IN COURT SERVICES

Association of Magistrates' Courts – 79 New Cavendish Street, London W1M 7RB. Tel: 0171 436 8524.

British Institute of Verbatim Reporters – 61 Carey Street, London WC2A 2JG.

Justices' Clerks' Society – The Law Courts, Petters Way, Yeovil, Somerset BA20 1SW. Tel: 01935 34159.

Magistrates' Courts Division – Training Branch, Lord Chancellor's Department, 4th Floor, Selborne House, 54–60 Victoria Street, London SW1E 6QB, Tel: 0171 210 8642. Can provide further information on openings within the Courts Service.

JUDGES AND MAGISTRATES

The Lord Chancellor's Department – Magistrates Court Division, Southside, 105 Victoria Street, London SW1E 6QT. Tel: 0171 210 3000. Can provide a career pack on working as a magistrate.

REGISTRAR OF BIRTHS, MARRIAGES & DEATHS
Society of Registration Officers – Karen Knapton, Honorary General Secretary, The Register Office, Aspen House (1st Floor), Temple Street, Swindon SN1 1SQ. Tel: 01793 522140 or 01793 521734.

Contact your local authority for further information.

Vacancies are sometimes advertised in the local press, and in such journals as *Opportunities*, *Local Government Chronicle*, *Municipal Journal* and *Local Government Review*. Copies of these should be available in public libraries.

CORONERS AND CORONERS' OFFICERS
Further details of the function of a coroner can be obtained from the local coroner's office. Contact your county constabulary for coroner's officer details; vacancies are usually advertised in the local press.

THE PRISON SERVICE
HM Prison Service – HQ, Room 425, Cleland House, Page Street, London SW1P 4LN. Tel: 0171 217 3000. Recruitment is dealt with at local level. Contact the governor of your local prison; the telephone number will be in *Yellow Pages*.

For information on the Accelerated Promotion Scheme, write to Room 418 at the above address, or phone 0171 217 6437.

Working in Police and Security, published by COIC.

SECURITY WORK
International Professional Security Association – IPSA House, 3 Dendy Road, Paignton, Devon TQ4 5DB. Tel: 01803 554849.
Security Industry Training Organisation Ltd – Security House, Barbourne Road, Worcester WR1 1RS. Tel: 01905 20004.

POLICE WORK
Police Recruiting Department – Room 516, Home Office, Queen Anne's Gate, London SW1H 9AT. Tel: 0171 273 3797.
Careers in the Police Force, published by Kogan Page.
The AGCAS booklet *Public Protection and Law Enforcement* includes graduate entry into the Police Force.
Working in Police and Security, published by COIC.

PRIVATE DETECTIVE WORK
Association of British Investigators – ABI House, 10 Bonner Hill Road, Kingston upon Thames, Surrey KT1 3EP. Tel: 0181 546 3368.
Institute of Professional Investigators – 31a Wellington Street, St. John's, Blackburn, Lancashire BB1 8AF. Tel: 01254 680072. (Can only assist people already working as professional investigators or those seeking information on NVQs and distance learning.)

FORENSIC SCIENCE
Forensic Science Service Headquarters – Priory House, Gooch Street North, Birmingham B5 6QQ.
Forensic Science Society – Clarke House, 18a Mount Parade, Harrogate HG1 1BX.

IMMIGRATION, CUSTOMS & EXCISE
HM Customs & Excise – London Recruitment Unit, 3rd Floor, Thomas Paine House, Angel Square, Torrens Street, London EC1V 1TA. Tel: 0171 865 3194.
Immigration and Nationality Department – Recruitment Section, Room 804, Apollo House, 36 Wellesley Road, Croydon CR9 3RR. Tel: 0181 760 8242.

PATENTS & TRADE MARKS
Chartered Institute of Patent Agents (CIPA) – Staple Inn Buildings, High Holborn, London WC1V 7PZ.
Tel: 0171 405 9450. (The CIPA will try to arrange for interested undergraduates to visit a patent agent to find out more.)

Institute of Trade Mark Agents – Canterbury House, 2–6 Sydenham Road, Croydon CR0 9XE. Tel: 0181 686 2052.

The Patent Office – (Recruitment), Cardiff Road, Newport, Gwent NP9 1RH. Tel: 01633 814544.

Patent work is included in *Information Management and Research* – an AGCAS leaflet -available for consultation in careers centres or from CSU, Despatch Department, Armstrong House, Oxford Road, Manchester M1 7ED. Tel: 0161 236 9816, ext 250/251.

Patent Agency – A Career for You? – free booklet from CIPA (see above).

A Career as a Patent Examiner – a free booklet from the Patent Office (see above).

COMPANY SECRETARY

Institute of Chartered Secretaries and Administrators – 16 Park Crescent, London W1N 4AH. Tel: 0171 580 4741.